Machine Learning for Sustainable Manufacturing in Industry 4.0

The book focuses on recent developments in the areas of error reduction, resource optimization, and revenue growth in sustainable manufacturing using machine learning. It presents the integration of smart technologies such as machine learning in the field of Industry 4.0 for better-quality products and efficient manufacturing methods.

- It focuses on machine learning applications in Industry 4.0 ecosystem, such as resource optimization, data analysis, and predictions.
- Highlights the importance of the explainable machine learning model in manufacturing processes.
- Presents the integration of machine learning and big data analytics from an Industry 4.0 perspective.
- Discusses advanced computational techniques for sustainable manufacturing.
- Examines environmental impacts of operations and supply chain from an industry 4.0 perspective.

This book provides scientific and technological insight into sustainable manufacturing by covering a wide range of machine learning applications such as fault detection, cyber-attack prediction, and inventory management. It further discusses resource optimization using machine learning in Industry 4.0 and explainable machine learning models for Industry 4.0. It will serve as an ideal reference text for senior undergraduate students, graduate students, and academic researchers in fields including mechanical engineering, manufacturing engineering, production engineering, aerospace engineering, and computer engineering.

Mathematical Engineering, Manufacturing, and Management Sciences

Series Editor: Mangey Ram, *Professor, Assistant Dean (International Affairs), Department of Mathematics, Graphic Era University, Dehradun, India*

The aim of this new book series is to publish the research studies and articles that bring up the latest development and research applied to mathematics and its applications in the manufacturing and management sciences areas. Mathematical tool and techniques are the strength of engineering sciences. They form the common foundation of all novel disciplines as engineering evolves and develops. The series will include a comprehensive range of applied mathematics and its application in engineering areas such as optimization techniques, mathematical modelling and simulation, stochastic processes and systems engineering, safety-critical system performance, system safety, system security, high assurance software architecture and design, mathematical modelling in environmental safety sciences, finite element methods, differential equations, reliability engineering, etc.

Biodegradable Composites for Packaging Applications
Edited by Arbind Prasad, Ashwani Kumar, and Kishor Kumar Gajrani

Computing and Stimulation for Engineers
Edited by Ziya Uddin, Mukesh Kumar Awasthi, Rishi Asthana, and Mangey Ram

Advanced Manufacturing Processes
Edited by Yashvir Singh, Nishant K. Singh, and Mangey Ram

Additive Manufacturing
Advanced Materials and Design Techniques
Pulak M. Pandey, Nishant K. Singh, and Yashvir Singh

Advances in Mathematical and Computational Modeling of Engineering Systems
Mukesh Kumar Awasthi, Maitri Verma, and Mangey Ram

Biowaste and Biomass in Biofuel Applications
Edited by Yashvir Singh, Vladimir Strezov, and Prateek Negi

For more information about this series, please visit: www.routledge.com/ Mathematical-Engineering-Manufacturing-and-Management-Sciences/ book-series/CRCMEMMS

Machine Learning for Sustainable Manufacturing in Industry 4.0

Concept, Concerns and Applications

Edited by
Raman Kumar
Sita Rani
Sehijpal Singh

CRC Press
Taylor & Francis Group
Boca Raton London New York

CRC Press is an imprint of the
Taylor & Francis Group, an **informa** business

First edition published 2024

by CRC Press
3825 Executive Center Drive, Suite 320, Boca Raton, FL 33431-8530

and by CRC Press
4 Park Square, Milton Park, Abingdon, Oxon, OX14 4RN

CRC Press is an imprint of Taylor & Francis Group, LLC

ISBN: 978-1-032-39305-6 (hbk)
ISBN: 978-1-032-59211-4 (pbk)
ISBN: 978-1-003-45356-7 (ebk)

DOI: 10.1201/9781003453567

Typeset in Sabon
by SPi Technologies India Pvt Ltd (Straive)

Contents

Preface

The manufacturing industry has rapidly transformed over the past few decades with the advent of Industry 4.0. This technological revolution has brought about significant changes in manufacturing processes, enabling them to become more efficient, cost-effective, and sustainable. Sustainable manufacturing is an essential component of Industry 4.0, which seeks to develop eco-friendly products using economical methods with optimal use of natural resources and energy.

Machine learning is a crucial tool that helps in achieving this goal. It has the potential to revolutionize the manufacturing industry by predicting future resource requirements, optimizing resource utilization, and facilitating efficient and eco-friendly manufacturing processes. Machine learning is a branch of artificial intelligence that enables computers to learn from data and improve their performance on specific tasks without being explicitly programmed.

This book highlights the concept, concerns, and applications of machine learning in sustainable manufacturing in Industry 4.0. The book covers machine learning and sustainable manufacturing, machine learning applications in Industry 4.0, resource optimization using machine learning, environmental impacts of operations and supply chains from Industry 4.0, and explainable machine learning models for Industry 4.0. In addition, this book provides several case studies that illustrate the application of machine learning in real-world domains of Industry 4.0 and smart factories.

The book's scope is to provide a comprehensive understanding of machine learning's potential in sustainable manufacturing and its role in the Industry 4.0 ecosystem. We believe this book will serve as a valuable resource for researchers, academicians, and experts in sustainable manufacturing and Industry 4.0. We thank all the contributors for their excellent work and hope readers will find this book insightful and informative.

Cheers to reading!
Raman Kumar, Sita Rani, Sehijpal Singh

Acknowledgments

We want to express our deepest gratitude to all the contributors who made this book possible. Their invaluable expertise, insights, and contributions have made this book an excellent resource for scholars, researchers, and experts in sustainable manufacturing and Industry 4.0. We would also like to thank the series editor, Dr. Mangey Ram, for his guidance, support, and encouragement throughout the book's development process.

Our sincere thanks go to the editorial team at the publishing house CRC Press (Taylor & Francis Group) for their support and cooperation in publishing this book.

Finally, we would like to acknowledge the efforts of our families, friends, and colleagues, who have provided us with encouragement and support throughout the writing process.

Once again, we express our heartfelt gratitude to all who contributed to this book.

Editor bios

Dr. Raman Kumar is an Assistant Professor in the Department of Mechanical and Production Engineering at Guru Nanak Dev Engineering College, Ludhiana, Punjab, India. He has completed his B.Tech (Mechanical Engineering), M.Tech (Production Engineering), and Ph.D. in Mechanical Engineering. He has five years of industry and 18 years of teaching and research experience.

His areas of interest are sustainable manufacturing, energy-efficient machining, optimization of processes, machine learning, and multi-criteria decision-making. He has taught at UG and PG levels, including Strength of Materials, Machining Sciences, Manufacturing Processes, Operation Research, and Industrial Automation and Robotics.

He has more than 100 research publications in national and international conferences and journals of repute. He is enlisted in the Stanford World list of the top 2% of Scientists of 2022 published by Elsevier. In addition, he acts as an Associate Editor of *Frontiers in Psychology*, *Frontiers in Public Health*, *Frontiers in Climate*, and *Frontiers in Rehabilitation Sciences*. He serves as a reviewer of the *Journal of Cleaner Production*, *Sustainable Cities and Society* (Elsevier), *International Journal of Precision Engineering and Manufacturing-Green Technology* (Springer), *Part B: Journal of Engineering Manufacture*, *Sustainability, Applied Sciences, Energies, Materials* (MDPI), *IEEE Sensor Journal, Frontiers*, etc.

He is a life member of ISTE and SAE India and has received ISTE Best Teacher Award-2021 (Section) and Faculty Excellence Award 2023 from the alumni association of GNDEC. He is a recipient of Best Paper Award at International Conference. He has participated in various faculty development programs, conferences, and expert lectures.

Scholar Profiles: Google Scholar; Research Gate; Publons; Orcid; Scopus; Vidwan; GNDEC Faculty

Dr. Sita Rani is a Faculty of Computer Science and Engineering at Guru Nanak Dev Engineering College, Ludhiana. She has completed her B.Tech and M.Tech degrees in Computer Science and Engineering from Guru Nanak Dev Engineering College, Ludhiana. She earned her Ph.D. in Computer Science and Engineering from I.K. Gujral Punjab Technical University, Kapurthala, Punjab, in 2018. She completed Postgraduate Certificate Program in Data Science and Machine Learning from the Indian Institute of Technology, Roorkee, in 2023. She is also Postdoctoral Research Fellow at South Ural State University, Russia, since May 2022. She has more than 20 years of teaching experience. She is an active member of ISTE, IEEE, and IAEngg. She is the receiver of the ISTE Section Best Teacher Award-2020 and International Young Scientist Award-2021. She has contributed to various research activities while publishing articles in renowned journals and conference proceedings. She has published six international patents and several books as author and editor. Dr. Rani has delivered many expert talks in AICTE-sponsored Faculty Development Programs and organized many International Conferences during her 20 years of teaching experience. She is a member of the Editorial Board and reviewer of many international journals of repute. Her research interests include parallel and distributed computing, machine learning, the Internet of Things (IoT), and healthcare.

Dr. Sehijpal Singh works as a Principal at Guru Nanak Dev Engineering College, Ludhiana, Punjab, India. He was Professor in the Mechanical and Production Engineering Department. He has received a Ph.D. in Mechanical Engineering from the Indian Institute of Technology, Roorkee, India. He has 26 years of teaching as well as research experience.

His areas of interest are non-conventional machining processes, energy-efficient machining, optimization of manufacturing processes, and decision-making. He taught subjects like Non-Conventional Machining, Metal Machining, and Manufacturing Processes.

He has guided 16 Ph.D. candidates. In addition, he has two patents, four books, and more than 100 research publications in National and International journals of repute, publications appearing in the *Journal of Cleaner Production, Materials and Manufacturing Processes, International Journal of Machine Tools and Manufacture,*

Journal of Materials Processing Technology, and *The International Journal of Advanced Manufacturing Technology*. In addition, he acts as a reviewer of various journals of Elsevier, Emerald, Springer, Inderscience, and SAGE and is a life member of ISTE, FIE, IIPE, and IIIE.

Scholar Profiles: Google Scholar; Vidwan; GNDEC Faculty; GNDEC Principal Desk

Contributors

Navneet Arora is a Professor in the Department of Mechanical and Industrial Engineering at the Indian Institute of Technology Roorkee, Uttarakhand, India. He completed his Bachelor's degree in Mechanical Engineering from REC Kurukshetra in 1988. He completed his Master's and Ph.D. from the University of Roorkee and Kurukshetra University in 1990 and 1997, respectively. His research interest is on simulation and modeling of manufacturing processes. He has published over 80 research papers in peer-reviewed international journals and has published/presented 64 papers at various national and international conferences.

Ajay Kumar Badhan is an Assistant Professor at Lovely Professional University in the Department of Computer Science and Engineering.

Abhishek Bhattacharjee is an Assistant Professor at Lovely Professional University in the Department of Computer Science and Engineering.

Harpreet Kaur Channi is an Assistant Professor at Chandigarh University, Gharuan, Mohali, Punjab, India. Her research interest mainly focuses on renewable energy resources and Sustainable Development Goals. She has published research papers in various conferences and journals. Her latest research is "A PV-Biomass off-grid Hybrid Renewable Energy System (HRES) for Rural Electrification: Design, Optimization and Techno-economic-environmental analysis" (2022).

Harnam Singh Farwaha is an Assistant Professor in Mechanical and Production Engineering Department at Guru Nanak Dev Engineering College, Ludhiana, Punjab, India. He received his Bachelor's and Master's from Guru Nanak Dev Engineering College, Ludhiana, and his Ph.D. from Punjabi University, Patiala, Punjab, India. His main research fields are non-conventional machining, automation, and robotics.

Rana Gill has done her Master's in Electronics and Communication Engineering with specialization in Image Processing and is currently pursuing Doctorate in ECE. Her area of interest is image processing, IOT. She has worked on government projects in field of implementation of IOT for smart agriculture. In addition to this she has guided number of students at graduate and postgraduate levels in field of image processing and IOT.

Chamkaur Jindal is an Assistant Professor in Mechanical and Production Engineering Department at Guru Nanak Dev Engineering College, Ludhiana, Punjab, India. He has completed his graduation and Master's in Mechanical Engineering from Punjabi University, Patiala, Punjab, India. His research interests focus on machine learning, surface engineering, materials characterization, etc. He has published research/review articles in reputed journals. Recently his research work for improving the service life of pulverized coal burner nozzle in boilers of thermal power plants has completed, and his Ph.D. thesis is submitted to IKG-PTU, Jalandhar, Punjab, India.

Ashima Kalra is working as an Assistant Professor in Electronics and Communication Department at Chandigarh Group of Colleges, Landran (Mohali), Punjab, India. Her research activities include designing model identification using neural networks, fuzzy systems, supervised learning, and machine learning. She has published more than 40 papers in reputed journals and three book chapters in Springer Series. She has published eight patents and four textbooks.

Rajwinder Kaur is pursuing Ph.D. in Digital Image Processing from Shri Guru Granth Sahib World University, Fatehgarh Sahib, India, and is currently working as an Assistant Professor at Baba Banda Singh Bahadur Engineering College, Fatehgarh Sahib. She has about 13 years of teaching experience. She has authored six publications in various journals (including *SCOPUS*) and has attended conferences as well.

Ranjodh Kaur is an Assistant Professor at Guru Nanak Dev Engineering College in the Department of Information Technology, Ludhiana. Her research interests focus on data mining and digital image processing. She has published her research work in three journals and one conference.

Sukhpreet Kaur Khalsa is working as a Data Center Specialist at the University of Alberta, Edmonton, AB, Canada. She earned her Master's in Computer Applications with a merit certificate from Punjab Agricultural University (Ludhiana, Punjab) and second Master's of Science in Internet working from the University of Alberta. She has spent time as a Ph.D. research scholar in the Department of Computer Applications at I.K. Gujral Punjab Technical University, Jalandhar, PB, India, as well. She has authored four publications in esteemed International refereed journals (including *SCI, SCOPUS, SCIe*) and reputed international conferences (IEEE, Springer, ACM) proceedings.

Raman Kumar is an Assistant Professor in the Department of Mechanical and Production Engineering at Guru Nanak Dev Engineering College,

Ludhiana, Punjab, India. He has completed his B.Tech (Mechanical Engineering), M.Tech (Production Engineering), and Ph.D. in Mechanical Engineering. He has five years of industry and 17 years of teaching and research experience. His areas of interest are sustainable manufacturing, energy-efficient machining, optimization of processes, machine learning, and multi-criteria decision-making.

Raman Kumar is a Professor at Chandigarh University in the Department of Mechanical Engineering. Raman Kumar holds a Bachelor's degree in Mechanical Engineering, Master's degree in Production Engineering, and Ph.D. in Mechanical Engineering from I.K. Gujral Punjab Technical University, Punjab, India. His research areas of interest include Industry 4.0, multi-attribute decision-making, and operation management.

Krishnendu Kundu is a Senior Principal Scientist at CSIR-CMERI-CoEFM, Ludhiana, Punjab. He completed his Ph.D. from GBPUAT Pantnagar in Agricultural Engineering. His research interests focus on Biofuel and Bioenergy. He has very vast experience in the area of biodiesel, biogas, and related biofuel technologies. Currently, he is working on various projects funded by various government agencies.

Aditi Mahajanan is a Research Scholar in the Department of Mechanical and Industrial Engineering at Indian Institute of Technology Roorkee, Uttarakhand, India. She completed her Bachelor's degree in Industrial and Production Engineering from Dr. B. R. Ambedkar National Institute of Technology, Jalandhar, in 2018. Her research interests focus on natural fiber-reinforced polymer composites, material and process selection, machine learning, and multicriteria decision-making. She has recently published in peer-reviewed international journals, such as *Composite Part C: Open Access* (2022), *Environment, Development and Sustainability* (2023), and *Materials Letters* (2023).

Sita Rani is a Faculty of Computer Science and Engineering at Guru Nanak Dev Engineering College, Ludhiana. She has served as Deputy Dean (Research) at Gulzar Group of Institutions, Khanna (Punjab). She has completed her B.Tech and M.Tech degrees in Computer Science and Engineering from Guru Nanak Dev Engineering College, Ludhiana. She earned her Ph.D. in Computer Science and Engineering from I.K. Gujral Punjab Technical University, Kapurthala, Punjab, in 2018. She has more than 20 years of teaching experience.

Pankaj Sawdatkar is Ph.D. student in Academy of Scientific and Innovative Research at CSIR-CMERI-CoEFM, Ludhiana, Punjab. He completed his Master's of Engineering from Pune University, Pune. His research interests focus on Advanced Manufacturing Processes and the application of AI and ML. He is currently working on the research topic of "Application of Machine Learning for Smart Machining."

Ramandeep Singh Sidhu is an Assistant professor at Guru Nanak Dev Engineering College, Ludhiana, Punjab, India, in the Department of Mechanical and production. His research interests focus on machine learning, friction stir welding, and carburization. His most research publications is *Joining of Dissimilar Al and Mg Metal Alloys by Friction Stir Welding* (2022).

Dilpreet Singh is Senior Scientist at CSIR-CMERI-CoEFM Ludhiana, Punjab. He completed his Ph.D. from IIT Delhi before joining CMERI. His research interests focus on biomedical engineering, advanced manufacturing processes, and additive manufacturing. Currently, he is working on various projects funded by various government agencies.

Inderdeep Singh is a Professor in the Department of Mechanical and Industrial Engineering at Indian Institute of Technology Roorkee, Uttarakhand, India. He completed his Bachelor's degree in Mechanical Engineering from NIT Hamirpur in 1998. He completed his Master's and Ph.D. from IIT Delhi in 2000 and 2004, respectively. His research interests focus on composite materials, conceptualization and development, manufacturing of polymer and metal matrix composites, and natural fiber-based green composites. He has published over 130 research papers in peer-reviewed international journals and published/presented 107 papers at various national and international conferences.

Sehijpal Singh works as a Principal at Guru Nanak Dev Engineering College, Ludhiana, Punjab, India. He was Professor in the Mechanical and Production Engineering Department. He has received a Ph.D. in Mechanical Engineering from the Indian Institute of Technology, Roorkee, India. He has 26 years of teaching as well as research experience. His areas of interest are non-conventional machining processes.

Gaurav Tewari is Governor Awardee and Founder of Gurukul Foundation. He is Chairperson of Aryabhatta Institute of Higher Research and Learning, Noida. He is working in several research activities. He has published 30+ research papers in reputed journals, four patents, and two books. His research areas are artificial intelligence, drone technology, microwaves, antennas, wireless communication, machine learning, IOT, and image processing.

Amit Verma is working at University Centre for Research and Development Department, Chandigarh University. His research area is artificial intelligence in the field of agriculture. He is currently working on the detection of the plant leaf diseases using the image processing and deep learning techniques. He works as Project Investigator in two DST-Funded projects related to artificial intelligence.

Chapter 1

Machine learning and sustainable manufacturing

Introduction, framework, and challenges

Ramandeep Singh Sidhu and Harnam Singh Farwaha

Guru Nanak Dev Engineering College, Ludhiana, India

1.1 INTRODUCTION OF MACHINE LEARNING

With extensive information and powerful computing, machine learning (ML) has developed to open up new possibilities for deciphering, measuring, and comprehending data-intensive systems in agricultural scenarios. Among other topics, ML is described as the branch of science that permits machines to acquire knowledge without being explicitly programmed. ML is employed in many scientific fields, including robotics, medicine, climatology, bioinformatics, biochemistry, meteorology, economic sciences, and food security (Li, Kermode, & De Vita, 2015).

Guidelines or suggestions on how to properly build ML algorithms can assist ensure accurate outcomes and predictions. Communities have established normative principles and best practices for scientific data management and the reproducibility of computer tools in biomedical research. However, there is a need in the ML community for a comprehensive set of guidelines that address the data, optimisation strategies, resultant model, and evaluation processes as a whole (Schütt et al., 2014).

Malware is one of the most potent hazards in the cyber domain despite the substantial advancement of cybersecurity mechanisms and their ongoing growth. Malware analysis uses techniques from various disciplines, including networking and programming evaluation, to examine malicious samples to gain a greater understanding of various topics, including their function and how they evolve (Gallagher et al., 2020). On the other hand, it is considerably harder to get around detection criteria that capture the semantics of a malicious sample since malware developers must make more intricate alterations. One of malware analysis's main objectives is finding more properties to strengthen security measures and make evasion as challenging as feasible. ML makes sense as a support for such a knowledge extraction method (Yang et al., 2022).

DOI: 10.1201/9781003453567-1

1.2 APPLICATION OF MACHINE LEARNING

1.2.1 Machine learning in materials science

ML models are more effective than people in comparison. They adopt a data-driven methodology and can analyse vast amounts of data without needing in-depth topic expertise or creative domain ideas. ML models can be customised. As inputs for ML models, raw goods characterisation information, such as spectrum and picture data, and conventional elements state parameters are also acceptable. ML models are precise as well. They have demonstrated outstanding forecast accuracy for various material properties at various scales. Modern ML techniques are precise, but their high predictability frequently comes at the cost of explainability. Typically, model reliability and model understandability are traded off. The most precise ML models (such as deep neural networks or DNNs) are typically challenging to understand and are frequently referred to as black boxes. The utility of ML models for common scientific activities, such as comprehending the hidden causal relationship, acquiring useful data, and developing new scientific ideas, has been constrained by this lack of explainability (Ranjan, Kumar, Kumar, Kaur, & Singh, 2022).

Many materials scientists distrust black-box ML models. Thus, concept definition is a crucial subject in explainable artificial intelligence (XAI) since accurate definitions assist in setting the discussion's context. Since model explanations might be deceptive, XAI also handles explanation evaluation. The use of XAI in scientific research is becoming more and more popular. Although some XAI methods are being used in materials ML investigations, most of the materials science community is still unaware of XAI (Gao et al., 2020). Applications of ML are shown in Figure 1.1.

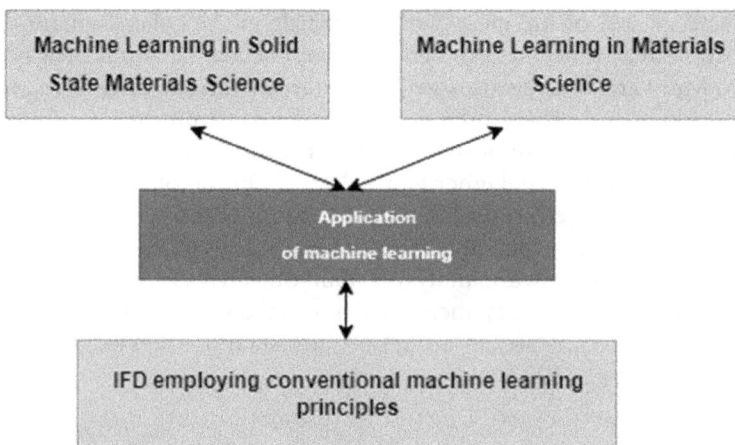

Figure 1.1 Application of machine learning.

1.2.2 Machine learning in solid-state materials science

In the past few years, there has been an unprecedented increase in interest in ML due to the accessibility of massive datasets, the improvements in algorithms, and the rise in computing power, which has been exponential. ML techniques are now successfully used to perform regression, classification, and clustering, particularly datasets with high dimensions. ML has shown extraordinary abilities in several areas, including go, self-driving cars, picture classification, etc. In the past, investigations were crucial to discovering and characterising novel materials. Because experimental research has such high resource and equipment needs, it must be carried out over a lengthy period for a very small number of materials. Due to these restrictions, serendipity or human intuition accounted for the most significant discoveries. Since ML techniques have recently been used in materials science, many reported solutions are fairly simple in architecture and sophistication. Several great evaluations of materials informatics and ML in materials science have covered this (Kumar, Rani, & Awadh, 2022). Due to the increase in the number of works utilising ML, however, a significant amount of research has already been published since the previous reviews, and the research landscape has quickly altered. Here, we focus on discussing and analysing the different uses of ML in materials of solid-state research, especially those that are most current. We briefly overview ML, aiming at the descriptors, principles, algorithms, and databases used in materials science. After that, we go over several uses of ML in the science of solid-state materials, including the identification of novel stable materials and the prediction of their structure, the properties measurement of materials using an ML algorithm, the development of ML functionals for the production of density functional theory (DFT) force fields for simulations in materials science. The active learning technique optimises the adaptive design process, improving an algorithm's interpretability (Loveland, Liu, Kailkhura, Hiszpanski, & Han, 2021).

1.2.3 IFD employing conventional ML principles

The link between monitoring data and machine health states, which has been a hot topic in machine health management, is pursued in great detail by fault diagnostics. The partnership has historically been anchored by engineers' vast expertise and depth of knowledge. For instance, a skilled engineer can identify engine issues based on anomalous sounds or pinpoint bearing faults by analysing vibration signals using cutting-edge signal processing techniques. Intelligent fault diagnosis (IFD) specifically intends to develop diagnosis models that can automatically establish a connection between the facts gathered and the health conditions of machines (Springenberg, Dosovitskiy, Brox, & Riedmiller, 2014).

The justification for using classic ML theories is discussed in this section. It also discusses IFD in the past using a regularly used diagnosis technique that comprises data collecting, artificial feature extraction, and health status detection. However, in engineering contexts, these solutions heavily rely on the specific knowledge that maintainers typically lack. The results of diagnosis employing signal processing techniques are also too complex for machine users to understand. Modern industrial applications, therefore, favour fault diagnosis techniques that can detect machine health conditions automatically. IFD hopes to accomplish the aforementioned goal with the aid of ML concepts. Machine defect evaluation has previously used several traditional ML concepts, like support vector machine (SVM) and artificial neural networks (ANN) (Kim et al., 2019).

1.3 ADVANTAGES AND DISADVANTAGES OF MACHINE LEARNING

1.3.1 Advantages of machine learning

- ML can evaluate a lot of data and spot certain trends and patterns that people might overlook. An e-commerce portal like Amazon, for example, can give its consumers proper products, coupons, and notifications by getting to know their web activity and purchasing patterns. It uses the information to display to them pertinent advertisements.
- Computers can generate forecasts and improve algorithms by themselves when they are empowered to learn. Antivirus software is a common example of this; it tends to prevent new hazards as they are discovered. ML is good at detecting spam.
- As they learn skills, ML algorithms keep advancing in both precision and effectiveness. Consider developing a model for weather forecasts. As your data collection grows, your algorithms get faster and more accurate at making predictions.
- Algorithms for ML are adept at handling multivariate and complex data in dynamic or unpredictable environments.

1.3.2 Disadvantages of machine learning

- Large, thorough, unbiased, and high-quality sets of data are necessary for ML training. Occasionally, they may need to wait until fresh information is created.
- The algorithms need sufficient time to develop and learn sufficiently to meet their objectives with a high level of relevance and accuracy for ML to be beneficial. Also, it uses a great deal of assets to function. You might then require your computer to have greater computing power.

- Another key difficulty is the ability to understand the information that the algorithms produce. Also, you must choose the strategies for your requirements thoroughly.
- Although being self-sufficient, ML is fallible. Think about developing an algorithm using insufficiently massive amounts of data. In the end, you get biased recommendations from a biased training dataset. Due to this, consumers see pointless advertisements. Such ML deficiencies can lead to a series of errors that may be unnoticed for a very long period. Also, it takes a bit to identify issues and even longer to find remedies.

1.4 FUTURE SCOPE OF MACHINE LEARNING

- One of the best employment options for the twenty-first century is ML. There are several high-paying career options there. Additionally, ML's potential for the future is poised to significantly alter the automation industry. Additionally, ML has several applications in India. So, if you want to contribute to the expanding digital world, you can build a successful career in the field of ML (Gramegna & Giudici, 2021).
- ML is automating tedious and repetitive activities, providing greater insights from data, and even enabling automobiles to drive themselves by giving robots the ability to "learn" to emulate human behaviour. The future of ML offers technologists far more and much more complex opportunities, even though the current state of ML is exciting as it is (Giles, Sengupta, Broderick, & Rajan, 2021).
- Statistical techniques are used in data mining projects to teach algorithms to produce classifications or predictions. They then affect crucial growth metrics by influencing company and application decision-making. In order to identify the most important business issues and the related data needed to address them, data scientists will be necessary, and demand for their services is anticipated to increase along with the growth of big data (Fullwood, Niezgoda, & Kalidindi, 2008).
- Artificial intelligence (AI) is a subcategory that helps software systems to increase their prediction accuracy without even being specifically created to do so. Old datasets are used by ML systems to foretell outcomes accurately. Some of the common applications of ML technology include spam filtering, fraud detection, smart healthcare systems, speech recognition, computer vision, and smart transportation (Shuo et al., 2021).

1.5 INTRODUCTION OF SUSTAINABILITY

Sustainable growth entails fulfilling our priorities without compromising the potential of future generations to do the same. Together with natural assets, we also need economic and social resources. Environmental advocacy is only one aspect of sustainable living. Several definitions of sustainability also take fairness and equality into account (Chandel, Kumar, & Kapoor, 2021). Although the idea of sustainability is a relatively recent one, the larger movement has roots in social justice, conservationism, globalism, and other long-standing causes. By the turn of the century, many of these ideas had come together, necessitating the term "sustainable development" (Fisher, 2011).

1.5.1 Applications of sustainable development

Figure 1.2 shows applications of sustainable development.

1.5.1.1 Analytics of big data for smart grid

Due to population growth and increasing urbanisation, energy consumption has significantly increased recently, which ultimately boosted carbon emissions. Around the world, several nations struggle to meet their residents' needs in key areas like public services. Early on in the development of the smart grid, extensive research was carried out to guarantee effective energy use and a sensible strategy for managing energy resources. By reducing energy intensity and its ensuing consequences on the environment, smart grid technologies are expected to improve the sustainability of the power supply (Roberts, 2009). However, it is still unknown how smart technology will interact with customer attitudes and behaviour in order to guarantee long-term sustainability. The social and technological facets of electricity supply are significantly impacted by the growth of smart grids. The social consequences of stakeholder impacts and the smart grid's resilience play a significant part in reaching the Sustainable Development Goals (SDGs) in the pursuit of technical developments (Channi & Kumar, 2022). In order to combine various sectoral programmes and create cogent cross-sectoral policies to explore options, it can be helpful to understand the relationships among the 17 SDG targets. When examining research papers on SDGs, the study "The Power of Data to Advance the SDGs," published by Elsevier and RELX SDG Resource Centre, was on SDGs. SDG 7 has received the majority of attention in the research, even if SDGs 12 and 6 are as important in achieving the sustainable goals. Thirteen per cent of the world's population, according to the report, lacks access to modern power. For their economic development, resource efficiency, and energy production, numerous emerging nations around the world, particularly China, India, Germany, the United Kingdom, and China, concentrate on the research that underpins SDG 7. In contrary to other SDGs, which only had a 3.5% compound annual growth rate (CAGR), the amount of research that went into SDG 7 increased by up to 9.1% from 2015. High-income producing nations are responsible for 57.2% of publications overall. China has contributed more research than any other country to SDG 7, or in other words, more than

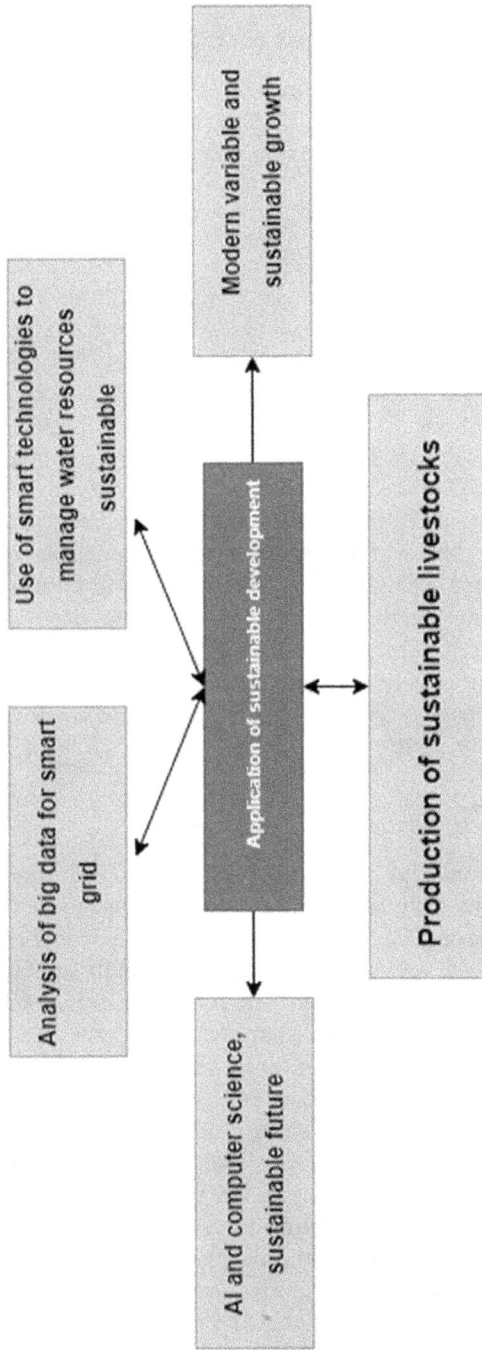

Figure 1.2 Applications of sustainable development.

167,700 papers (57.2%), whereas India has the fewest publications. About 23% of the SDGs' research is collaborative; however, SDG 7 research has an annual increase in its Field-Weighted Citation Impact (FWCI) (Mackenzie, 2007).

1.5.1.2 Use of smart technologies to manage water resources sustainably

Due to factors such as global population growth, climate change, and increased industrialisation, resources for freshwater have been declining at a concerning rate. Researchers estimate that 52% of the world's population will reside in water-scarce or dry regions by the year 2050. This makes the development and execution of environmentally friendly water management policies necessary due to issues with public health, the environment, and society. The SDGs of the United Nations have established explicit targets for achieving universal – by 2030, ensuring everyone has adequate access to safe drinking water (SDG #6) and ensuring that freshwater is used responsibly (SDG #12). The world's population will be 68% urban by 2050, according to predictions. Urban water shortages have been accelerated by growing urbanisation, creating serious water imbalances and shortages. Therefore, residential and municipal levels of sustainable activities and technology towards urban water governance have been emerging as essential. The fourth wave of industrialisation has also led to a "smart" transition in urban water management, and the only realistic strategy to ensure water sustainability in the cities of the future is this one (Gupta, Reynolds, & Patel, 2010).

1.5.1.3 Production of sustainable livestock

Population expansion and shifting consumption patterns endanger the viability of numerous crop and livestock production systems in numerous nations. Increased cropping, modifications to the pattern and intensity of cropping, and excessive grazing on rangelands result in soil deterioration, which puts food production at risk. The livestock industry in particular is frequently implicated. To fulfil the rising need for food and jobs, as well as to stop the environment from degrading, animal development is required. A change in one function might have an impact on other functions, which highlights the necessity for an organisational approach in the development of sustainable livestock. Many of the limitations stated in the preceding paragraphs are now acknowledged, and in some situations, this has caused the policy to be reoriented. A farming systems approach has been adopted by non-governmental organisations (NGOs) and by institutions of research and international development organisations, putting an emphasis on the socioeconomic study of the entire farm and the farmers' participation in the development and research process. There is agreement that interventions must be viewed in the context of optimal resource development, aiming to restore and maintain soil fertility. This contrasts with conventional, unsustainable industrial strategies that encourage land expansion or mining. The goal for both industrialised and emerging nations is to stop the current scenario of environmental deterioration and to achieve enhancements in output that are sustainable. External inputs must be used more effectively even if they are necessary. The Brundtland Committee noted that "more needs to be done by less (Bonabeau, 2002).

1.5.1.4 Modern variations and sustainable growth

The first modern grain and rice variants launched at this time by the newly founded research centres of international agricultural were heralded as miracle seeds and the key to agricultural expansion and advancement; by the mid-1960s, the world's population had doubled, population growth was outpacing enhanced production of food, and there wasn't much undeveloped area to expand farming on. These modern varieties (MVs), which included one-third to half of Asia's rice and wheat fields area in Asia by the early 1970s, increased yields by up to 50%, and its enormous potential was a result of the adoption of contemporary plant breeding methods to make them extremely yield-responsive to the usage of outside inputs such as chemical fertiliser, water for irrigation, and pesticides. Despite the fact that this was a technical achievement, questions concerning the larger implications of MVs soon started to be raised (Peters, Baumann, Zimmermann, Braun, & Weil, 2017).

1.5.1.5 AI and computer science for a sustainable future

In the past few years, interest in AI and sustainability has greatly increased. This is somewhat in line with the rising popularity of computing in general and sustainability in particular. However, rather than merely following this bigger movement, AI is actively driving it. In fact, it's difficult to conceive a scenario in which AI is not essential to comprehending and controlling the enormous intricacy of preserving a healthy environment in the face of widespread and transformational anthropogenic. The founding of the Computing Environmental Institute, which focuses on AI and several environmental consequences such as ecology and clean technology, is a notable and technically relevant milestone in the development of AI and environment. The institute swiftly drew in additional researchers, educators, government officials, and business leaders. In 2011, the Society for the Development of Artificial Intelligence conference created a specific session on computational ecology as a result of the first conference on the topic, which was held in 2009, followed by two more in 2010 and 2011 (Kotsiopoulos, Sarigiannidis, Ioannidis, & Tzovaras, 2021).

The computer fields mentioned previously encourage the use of AI techniques to speed up development. For instance, scheduling and planning concerns throughout the manufacturing process and automated monitoring during the usage stages such as condensing distribution networks to lessen environmental impact are both present in green IT. Another emerging trend is AI for ecological sustainability such as wrap design, which aims to avoid waste using low-energy recycling methods. Another key sustainable methodology is ML. The use of ML methods ranges from learning to count and identify individuals to estimating the distributions of a certain species. Discovering usage trends for various appliances from basic home sensors and discovering ways to foresee breakdowns in aged civic infrastructure. Learning is crucial to achieving the huge promise of computing for customisation, which makes it feasible to characterise people in complex ways that go much beyond the data typeviewpoint labels of "liberal," "conservative," "wasteful," or "thrifty." With these more detailed descriptions, we can better connect individual activities to sustainability, leading to significant waste reduction and energy savings for instance in convenience services. Think about leaving hotel air conditioners on to prevent a new guest from feeling uncomfortable for a few minutes. The air conditioners in the area assigned specifically for me will be turned off in an interconnected ubiquitous computing system that has learnt my interests at least until arrive (Jamil, Iqbal, Ahmad, & Kim, 2021).

1.5.2 Advantages and disadvantages of sustainability

1.5.2.1 Advantages of sustainability

- Sustainability protects the ecosystem's health and biocapacity.
- Sustainability promotes the happiness of people and communities. Less wastage and pollutants, reduced emissions, more jobs, and a more equitable share of wealth are all promoted through an improved sustainable economy.
- Many customers now consider a company's environmental impact before buying something. Consumers seem to be more willing to back companies that show a commitment to protecting the natural world. Consumers enjoy, appreciate, and spend more at companies that give back to their neighbourhoods. Due to this, many well-known corporations are placing more focus on the promotion of their charity involvement in the community in their advertising campaigns, and it works.
- One of the easiest and most efficient methods to make a business profitable is to use sustainable waste management strategies. The practice of decreasing waste is now standard in any sustainable organisation, going all the way back to the early attempts made by businesses to separate out materials for recycling nearly three decades ago to the zero-waste projects being pursued today (Chandel et al., 2021; Ranjan et al., 2022).

1.5.2.2 Disadvantages of sustainability

- If you're not careful, using ecologically friendly procedures will result in greater operating costs because they demand more expensive tools and materials.
- It can be challenging to achieve sustainable development if there are those who disagree with this type of strategy since it needs society and governments participating in the production process to be committed to using only environmentally friendly products and procedures.
- Jobs can be created through sustainable development for people who want them, but there is also a chance that some industries could contract or possibly completely vanish due to competition from newcomers whose firms are based on sustainability principles rather than just profit margins (such as renewable energy providers). In some industries, this can result in job losses.
- For individuals who think these accusations are unjustified or even harmful for society as a whole, it can be distressing because sustainable development is sometimes criticised for being too idealistic or impractical, or for not placing enough emphasis on profit motives or economic progress.

1.5.3 Future scope

- Countries all across the world have taken steps to promote sustainability, and the companies that operate there can have a significant impact. To reduce food waste, a regulation in France requires stores to donate leftovers to charitable organisations. To avoid accumulation of excess waste and recycling waste in landfills, Switzerland has invented a technique to convert them into energy. By 2025, Copenhagen, Denmark, will achieve CO_2-neutral status as the world's first metropolis. The viability of a nation's economy is a key factor in its sustainability. Every action has an effect on overall sustainability, thus any sustainability initiatives businesses implement will have a positive impact on the environment in which we live (Wang, Chen, Hong, & Kang, 2018).
- Looking at global concerns about sustainable environmental rank among the top issues, sustainability of habitats, ecosystems, and the earth as a whole are important issues. With the help of biotechnological systems, tools, and processes, future development can be shaped to preserve the sustainability of our environment (Leszczyna, 2018).
- Sustainability includes cultural, ethical, technological, socioeconomic, legal, and last but not the least environmental aspects. All businesses may identify activities to support John Elkington's triple bottom line: people, profit, and planet. This is true whether they are high-end fashion houses or overvalued internet startups. Although profit has

dominated the triad throughout the majority of this planet's illustrious history, earlier civilisations did not view sustainability and profit as mutually exclusive. You should visit Rome, Paris, or Ephesus to behold once-profitable economies that were designed to last forever. It will be brought to your attention that, in reality, profit was never the bad guy, but that impromptu swiftness for the sake of profit. Hurricanes' devastating effects on affordable housing, quick "solutions" like redlining, and how "fast food" has led to many people fighting protracted health battles are just a few examples. We not only misunderstand the extent of sustainability and the effects it has on many societal aspects, but we also believe that quick and "tokenised" greenwashing is an almost humorous method to get around the issue. Change is urgently needed in our environment, schools, infrastructure, energy, transportation, culture, etc., but any solution must be long-lasting (Gifty, Bharathi, & Krishnakumar, 2019).

1.6 INTRODUCTION OF INDUSTRY 4.0

Three industrial changes have occurred since the 1700s. The very first industrial changes (Industry 1.0) took place between the late 1700s and the middle of the 1800s, as businesses shifted away from using water and wind to power production machines and embraced the use of steam power and the steam engine more and more. The ability to carry products and materials was boosted by steam-powered trains and ships (Mason & Grijalva, 2019). Other notable inventions were also made during this time period, such as gas lighting, which allowed companies to stay open later at night than they could have with candles or oil lights. Between 1870 and 1914, there was a second industrial change, sometimes known as Industry 2.0. The development of efficient steel production techniques sparked a significant railway network expansion and produced more dependable machinery, both of which increased the industrial output (Kathirgamanathan, De Rosa, Mangina, & Finn, 2021). The invention of electricity revolutionised manufacturing because electrical motors and machinery were more adaptable, effective, and simple to maintain than steam-powered technology. Huge capacity and productivity gains were made possible by the introduction of factory assembly lines and mass production techniques (Antonopoulos et al., 2020). Automobiles made it simpler for businesses to acquire labour and deliver commodities to places the rails did not go. Improved communication tools like the telegraph, telephone, and radio allowed businesses to react to market demands more quickly. The latter half of the twentieth century and the early years of the current one saw the third industrial change (Industry 3.0). Computer technology and the digital age experienced a remarkable rise during this time. Because of developments in computer technology, the same processing capability that formerly required several floors of wall-to-wall computers in the 1970s may now fit in the palm of your hand. The creation of the internet brought people together and created new possibilities for the exchange and utilisation of knowledge and data. They were termed industrial "revolutions" because the invention behind them significantly transformed how goods were produced and how work was done, not merely slightly enhancing productivity and efficiency (Gururajapathy, Mokhlis, & Illias, 2017).

Now the globe is currently experiencing the fourth industrial change, also known as Industry 4.0. According to Daniel Burrus, "The emergence of digital industrial technologies is the general description of Industry 4.0 (Duchesne, Karangelos, & Wehenkel, 2020). With the advent of Industry 4.0, mankind may collaborate with machines in novel, extremely productive ways." Smart technologies are being used by Industry 4.0 to revolutionise supply chain automation, supervision, and assessment. Industrial Internet of Things (IIoT) and intelligent cyber-physical technologies are appropriate in different situations. The monitoring and control of physical items like equipment, robotic systems, and automobiles use computer-based technologies. This is the core of Industry 4.0 (Hastie, Tibshirani, Friedman, & Friedman, 2009).

1.6.1 Nine pillars of Industry 4.0 technologies

Industry 4.0 is focused on nine fundamental aspects. These innovations enable the development of intelligent, auto systems by fusing the actual world with the digital one. Supply chains now use certain types of material removal techniques and businesses (Huo, Bouffard, & Joós, 2021). When they are combined, Industry 4.0 is fully realised. Figure 1.3 represent nine pillars of Industry 4.0 (Kumar et al., 2022).

- **Big data and AI analytics:** Industry 4.0 collects big data from various sources, including manufacturing machinery and Internet of Things (IoT) gadgets, enterprise resource planning (ERP) and customer relationship management (CRM) systems, and forecast and transportation apps. Analytics powered by AI and ML are applied to the data in real time – and insights are leveraged to improve decision-making and automation in every area of supply chain management: supply chain planning, logistics management, manufacturing, R&D and engineering, enterprise asset management (EAM), and procurement (Ugurlu, Oksuz, & Tas, 2018).
- **Horizontal and vertical integration:** Integrating vertically and horizontally is the foundation of Industry 4.0. The "field level" of horizontal integration refers to the production floor, various production facilities, and the entire supply chain, where operations are tightly connected. With vertical integration, all organisational levels are connected, allowing for seamless data flow from the shop floor to the top floor

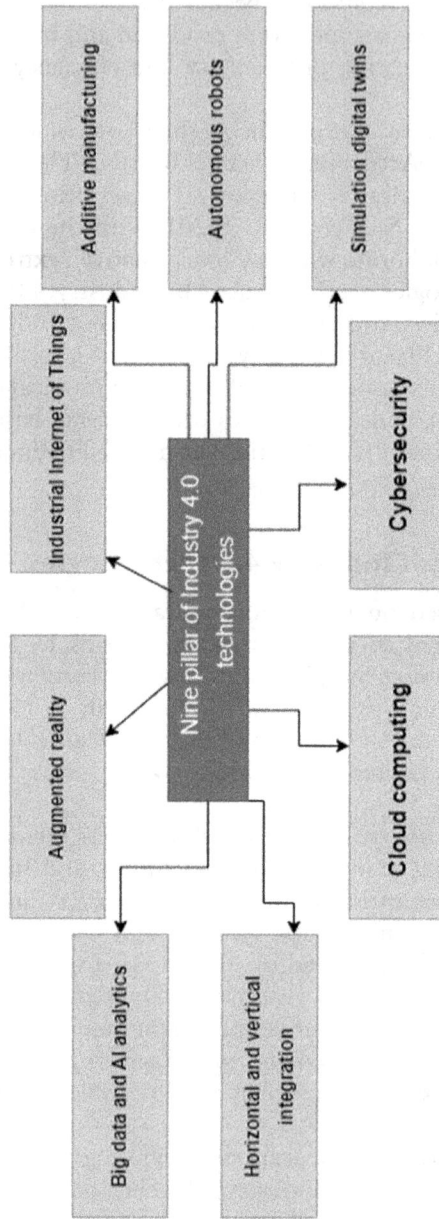

Figure 1.3 Nine pillars of Industry 4.0.

and back again. In other words, data and information boundaries are no longer an issue because manufacturing is intimately integrated with corporate activities like R&D, quality assurance, marketing and sales, and other departments (Yildiz, Bilbao, & Sproul, 2017).

- **Cloud computing:** The "great enabler" of Industry 4.0 and technological advancement is cloud computing. The capabilities of modern cloud computing go far beyond scalability, cost effectiveness, and speed. It lays the groundwork for the most cutting-edge technologies, from AI and ML to the Internet of Things, and it gives companies the tools to innovate (Kumar, Banga, & Kaur, 2020). The cyber-physical systems at the centre of Industry 4.0 use the cloud for communication and coordination, and the data that powers these technologies is stored there (Habib, Kamran, & Rashid, 2015).

- **Augmented reality (AR):** Industry 4.0's central idea is enhanced reality, which superimposes digital content on a real-world setting. With an AR system, workers can view real-time IoT data, digital parts, maintenance or assembly instructions, training materials, and more while gazing at a physical object, such as a piece of machinery or a product, using smart glasses or mobile devices. Although AR is still in its infancy, it has significant implications for technician safety, training, and quality assurance as well as maintenance, service, and assurance (Cao et al., 2020).

- **Industrial Internet of Things (IIoT):** The Internet of Things is absolutely essential to Industry 4.0 and more particularly, the IIoT – the two names are frequently used alternatively. In Industry 4.0, the majority of physical objects – devices, robots, machinery, equipment, and products – use sensors and Radio-Frequency Identification (RFID) tags to transmit real-time information about their state, functionality, or location. Companies may use this technology to manage supply chains more efficiently, design and adapt products more quickly, prevent equipment breakdowns, remain abreast of consumer tastes, track products and inventories, and do a lot more (François-Lavet, Taralla, Ernst, & Fonteneau, 2016).

- **Additive manufacturing/3D printing:** Another significant technology powering Industry 4.0 is additive manufacturing, sometimes known as 3D printing. Rapid prototyping was the original purpose of 3D printing, but it now has a wider range of uses, including mass customisation and dispersed manufacturing (Singh et al., 2021). For instance, using 3D printing, products and parts can be kept in virtual inventories as design files and printed when needed, cutting down on travel expenses and distance (Koirala, van Oost, & van der Windt, 2018).

- **Autonomous robots:** A new group of intelligent machines is emerging with Industry 4.0. performed by a programme. A new generation of autonomous robots is emerging with Industry 4.0. Autonomous robots, which are designed to carry out activities with little to no human assistance, come in a wide range of shapes and sizes, from inventory scanning drones to mobile autonomous robots used in pick-and-place operations (Chodha, Dubey, Kumar, Singh, & Kaur, 2021). These robots are capable of carrying out tough and sensitive jobs and can detect, interpret, and act on information they receive from their environment, thanks to advanced software, AI, sensors, and machine vision (Du & Li, 2020).

- **Simulation/digital twins:** Based on data from Internet of Things (IoT) sensors, a digital twin is a virtual imitation of a physical machine, product, process, or system. Businesses may better comprehend, examine, and enhance the functionality and upkeep of industrial devices and systems, thanks to this key element of Industry 4.0. For instance, an asset operator can utilise a digital twin to pinpoint a specific failing component, foresee foreseeable problems, and increase uptime (Koirala et al., 2018).
- **Cybersecurity:** Effective cybersecurity is essential, given the heightened connectivity and big data utilisation in Industry 4.0. Companies may automate threat detection, protection, and response – and lower the risk of data breaches and production lags across their networks – by deploying a Zero Trust infrastructure and techniques like ML and blockchain (Cao et al., 2020).

1.6.2 Benefits and drawbacks of industrial manufacturing

1.6.2.1 Benefits of industrial manufacturing

- **Intelligent products:** Create conscience, connected objects that can exchange information about their position, past transactions, storage conditions, and more. With the aid of the data these devices give, you can improve anything from customer satisfaction and quality of products to transportation and R&D. Additionally, they may anticipate service needs, obtain improvements virtually, and open the door for novel client marketing strategies (Pan et al., 2020).
- **Intelligent factories:** Run "smart" factories – highly automated, digitally enhanced workplaces that make the most of cutting-edge tools like big data, AI, robots, and the IoT. These factories use self-correcting smart manufacturing 4.0 processes and are also referred to as "Factories 4.0" and enable the efficient and widespread delivery of bespoke products (Pan et al., 2020).
- **Intelligent assets:** Nearly all of today's operational tangible infrastructure have established sensors, and when combined with data analysis and the IoT, they completely revolutionise enterprise asset management. Professionals may tightly link assets and business operations

with smart resources, use adaptive and prediction maintenance, anticipate and avoid delays, and keep an eye on resource performance in a timely manner (Pan et al., 2020).

- **Empowered people:** Your systems will always necessitate people, regardless of how autonomous they get, and give them access to live sensor data and technologies like AI so they can stay informed about what's happening on the work floor and be prepared to respond quickly to problems as they arise. They can also use augmented reality apps and wearable technology to solve problems, monitor their health, and stay safe (Huo et al., 2021).

1.6.2.2 Disadvantages of Industry 4.0

- This process requires a large initial investment.
- Technology requires companies to constantly update, which is unsustainable for most organisations.
- Industry 4.0 will create a gap between companies that have adapted and those that haven't.
- Specialised personnel are needed to analyse and monitor automated processes (Chung, Maharjan, Zhang, & Eliassen, 2021).

1.6.3 Steps to transform traditional industry into Industry 4.0

- Make a data model that depicts how you should manage your firm.
- Redesign your business procedures and search for more productive working methods.
- Increase your business insights by utilising your corporate data.
- Start producing business results from your data using low-code applications.
- Make outcome-based software that can be swiftly released in sprints (Gong & Ruan, 2020).

1.7 RELATIONSHIP BETWEEN INDUSTRY 4.0, MACHINE LEARNING, AND SUSTAINABLE MANUFACTURING SYSTEMS

1.7.1 Effect of machine learning on Industry 4.0

The old manufacturing period is evolving into the smart manufacturing era of Industry 4.0 due to ML. New opportunities are being created by this paradigm change. The Internet of Things (IoT) is a subset of the Industrial Internet of Things (IIoT), which is a network of physical items that are digitally connected and enable data exchange and communication through the internet. Through a variety of sensors, RFID tags, software, and electronics that are linked to machines to capture real-time data, IIoT connects machines. A tremendous amount of industrial data-rich surroundings relating to all facets of production may be created and stored, thanks to the proliferation of smart sensors and the Internet of Things. Based on information obtained from the IIoT, digital twin performs an online simulation. A digital twin is a representation of a physical good, machine, process, or system that enables businesses to more fully comprehend, analyse, and improve their operations through real-time simulations (Balouji, Bäckström, & Hovila, 2020).

- Data analytics can improve assembly-line efficiency.
- A better customer experience, including personalised value propositions and personalisation.
- Managing inventory through (a) real-time knowledge and visibility of inventories through supply lines and (b) improvements in delivery routes.
- To provide real-time asset management and to reduce loss from delayed, damaged, or lost products in transit, real-time notifications should be used.
- Using (a) analytics-backed simulations and (b) product modelling helps decrease errors, and the adjustments made during product creation enhance motivation of workers and packaging.
- Predictive maintenance increases the lifespan of assets. (a) Asset Management (b) Increasing the availability of assets. (c) Finding errors and flaws. (d) The avoidance of unanticipated downtimes.
- Using location-based IoT services improves supply chain visibility with actionable data.

1.7.2 Effect of sustainability manufacturing on Industry 4.0

Depending on the goal and application, sustainable manufacturing is viewed from a variety of angles; nevertheless, among industrial participants, the angles that are most frequently acknowledged are surroundings, culture, finance, technologies, and quality management. The foundational elements of environmental sustainability – environment, society, and economy – among the aforementioned dimensions are frequently referred to as the "triple bottom line" (TBL).

In order to be sustainable, it is necessary to create manufacturing processes and products that are recyclable and have no adverse effects on the environment. While technological advancements aid in the creation of novel procedures and goods, sustainable growth requires integration of digital technology. Companies are actively focusing on this convergence in order to fully profit from Industry 4.0, which is sustainable production (Lu et al., 2019).

1.8 CONCLUSIONS

Industries utilising technology that enables them to anticipate bad and incorrect behaviour optimise production procedures and thoroughly analyse the market or demand in order to better understand it and subsequently respond to client requests. All of this is made possible by the various uses of ML.

Manufacturers today must prioritise sustainability as a top commercial concern in addition to their social responsibilities. Manufacturing companies may promote sustainability along the whole value chain with the aid of Industry 4.0 technology, advancing both their own and the interests of their stakeholders.

1.8.1 Future scope of machine learning, sustainable manufacturing in Industry 4.0

One of the best employment options for the twenty-first century is ML. There are several high-paying career options there. Additionally, ML's potential for the future is poised to significantly alter the automation industry. Additionally, ML has several applications in India. So, if you want to contribute to the expanding digital world, you can build a successful career in the area of ML. We shall talk about numerous trends and the potential of ML in this blog.

The energy and utilities sector is already ahead of other sectors when it comes to sustainable practices, maybe as a result of greater regulatory scrutiny. Sustainable decisions will usher in an exciting new era for the energy and utilities sector as the cost of using renewable energy continues to decline. The most difficult problems facing your business today may be resolved by Industry 4.0, but it's crucial to think further and more broadly than that. Plan your AI innovation around the specific objectives and benefits of your business, and consider what the future might include when AI is widely used.

REFERENCES

Antonopoulos, I., Robu, V., Couraud, B., Kirli, D., Norbu, S., Kiprakis, A., … Wattam, S. (2020). Artificial intelligence and machine learning approaches to energy demand-side response: A systematic review. *Renewable and Sustainable Energy Reviews*, *130*, 109899.

Balouji, E., Bäckström, K., & Hovila, P. (2020, October 26–28). *A Deep Learning Approach to Earth Fault Classification and Source Localization. Paper presented at the 2020 IEEE PES Innovative Smart Grid Technologies Europe (ISGT-Europe)*.

Bonabeau, E. (2002). Agent-based modeling: Methods and techniques for simulating human systems. *Proceedings of the national academy of sciences*, *99*(suppl_3), 7280–7287.

Cao, D., Hu, W., Zhao, J., Zhang, G., Zhang, B., Liu, Z., ... Blaabjerg, F. (2020). Reinforcement learning and its applications in modern power and energy systems: A review. *Journal of Modern Power Systems and Clean Energy, 8*(6), 1029–1042. doi:10.35833/MPCE.2020.000552

Chandel, R. S., Kumar, R., & Kapoor, J. (2021). Sustainability aspects of machining operations: A summary of concepts. *2nd International Conference on Functional Materials, Manufacturing and Performances, ICFMMP 2021, 50,* 716–727. doi:10.1016/j.matpr.2021.04.624

Channi, H. K., & Kumar, R. (2022). The role of smart sensors in smart city. In: *Vol. 92. Studies in big data* (pp. 27–48): Springer Science and Business Media Deutschland GmbH.

Chodha, V., Dubey, R., Kumar, R., Singh, S., & Kaur, S. (2021). Selection of industrial arc welding robot with TOPSIS and Entropy MCDM techniques. *2nd International Conference on Functional Materials, Manufacturing and Performances, ICFMMP 2021, 50,* 709–715. doi:10.1016/j.matpr.2021.04.487

Chung, H. M., Maharjan, S., Zhang, Y., & Eliassen, F. (2021). Distributed deep reinforcement learning for intelligent load scheduling in residential smart grids. *IEEE Transactions on Industrial Informatics, 17*(4), 2752–2763. doi:10.1109/TII.2020.3007167

Du, Y., & Li, F. (2020). Intelligent multi-microgrid energy management based on deep neural network and model-free reinforcement learning. *IEEE Transactions on Smart Grid, 11*(2), 1066–1076. doi:10.1109/TSG.2019.2930299

Duchesne, L., Karangelos, E., & Wehenkel, L. (2020). Recent developments in machine learning for energy systems reliability management. *Proceedings of the IEEE, 108*(9), 1656–1676. doi:10.1109/JPROC.2020.2988715

Fisher, D. H. (2011). Computing and AI for a sustainable future. *IEEE Intelligent Systems, 26*(6), 14–18.

François-Lavet, V., Taralla, D., Ernst, D., & Fonteneau, R. (2016). *Deep reinforcement learning solutions for energy microgrids management. Paper presented at the European Workshop on Reinforcement Learning (EWRL 2016).*

Fullwood, D. T., Niezgoda, S. R., & Kalidindi, S. R. (2008). Microstructure reconstructions from 2-point statistics using phase-recovery algorithms. *Acta Materialia, 56*(5), 942–948.

Gallagher, B., Rever, M., Loveland, D., Mundhenk, T. N., Beauchamp, B., Robertson, E., ... Han, T. Y.-J. (2020). Predicting compressive strength of consolidated molecular solids using computer vision and deep learning. *Materials & Design, 190,* 108541.

Gao, S.-H., Tan, Y.-Q., Cheng, M.-M., Lu, C., Chen, Y., & Yan, S. (2020, August 23–28). *Highly efficient salient object detection with 100k parameters. Paper presented at the Computer Vision–ECCV 2020: 16th European Conference,* Glasgow, UK, Proceedings, Part VI.

Gifty, R., Bharathi, R., & Krishnakumar, P. (2019). Privacy and security of big data in cyber physical systems using Weibull distribution-based intrusion detection. *Neural Computing and Applications, 31*(Suppl 1), 23–34.

Giles, S. A., Sengupta, D., Broderick, S. R., & Rajan, K. (2021). Machine-learning-based intelligent framework for discovering refractory high-entropy alloys with improved high-temperature yield strength. *arXiv preprint arXiv:2112.02587.*

Gong, R., & Ruan, T. (2020). A new convolutional network structure for power quality disturbance identification and classification in micro-grids. *IEEE Access, 8,* 88801–88814. doi:10.1109/ACCESS.2020.2993202

Gramegna, A., & Giudici, P. (2021). SHAP and LIME: An evaluation of discriminative power in credit risk. *Frontiers in Artificial Intelligence, 4*, 752558.

Gupta, S., Reynolds, M. S., & Patel, S. N. (2010). *ElectriSense: single-point sensing using EMI for electrical event detection and classification in the home. Paper presented at the Proceedings of the 12th ACM international conference on Ubiquitous computing.*

Gururajapathy, S. S., Mokhlis, H., & Illias, H. A. (2017). Fault location and detection techniques in power distribution systems with distributed generation: A review. *Renewable and Sustainable Energy Reviews, 74*, 949–958. doi:10.1016/j.rser.2017.03.021

Habib, S., Kamran, M., & Rashid, U. (2015). Impact analysis of vehicle-to-grid technology and charging strategies of electric vehicles on distribution networks – A review. *Journal of Power Sources, 277*, 205–214. doi:10.1016/j.jpowsour.2014.12.020

Hastie, T., Tibshirani, R., Friedman, J. H., & Friedman, J. H. (2009). *The elements of statistical learning: data mining, inference, and prediction* (Vol. 2): Springer.

Huo, Y., Bouffard, F., & Joós, G. (2021). Decision tree-based optimization for flexibility management for sustainable energy microgrids. *Applied Energy, 290*, 116772. doi:10.1016/j.apenergy.2021.116772

Jamil, F., Iqbal, N., Ahmad, S., & Kim, D. (2021). Peer-to-peer energy trading mechanism based on blockchain and machine learning for sustainable electrical power supply in smart grid. *IEEE Access, 9*, 39193–39217.

Kathirgamanathan, A., De Rosa, M., Mangina, E., & Finn, D. P. (2021). Data-driven predictive control for unlocking building energy flexibility: A review. *Renewable and Sustainable Energy Reviews, 135*, 110120. doi:10.1016/j.rser.2020.110120

Kim, B., Seo, J., Jeon, S., Koo, J., Choe, J., & Jeon, T. (2019). *Why are saliency maps noisy? cause of and solution to noisy saliency maps. Paper presented at the 2019 IEEE/CVF International Conference on Computer Vision Workshop (ICCVW).*

Koirala, B. P., van Oost, E., & van der Windt, H. (2018). Community energy storage: A responsible innovation towards a sustainable energy system? *Applied Energy, 231*, 570–585. doi:10.1016/j.apenergy.2018.09.163

Kotsiopoulos, T., Sarigiannidis, P., Ioannidis, D., & Tzovaras, D. (2021). Machine learning and deep learning in smart manufacturing: The Smart Grid paradigm. *Computer Science Review, 40*, 100341. doi:10.1016/j.cosrev.2020.100341

Kumar, R., Banga, H. K., & Kaur, H. (2020). *Internet of Things-Supported Smart City Platform. Paper presented at the 1st International Conference on Sustainable Infrastructure with Smart Technology for Energy and Environmental Management, FIC-SISTEEM 2020.*

Kumar, R., Rani, S., & Awadh, M. A. (2022). Exploring the application sphere of the internet of things in Industry 4.0: A review, Bibliometric and Content Analysis. *Sensors, 22* (11). doi:10.3390/s22114276

Leszczyna, R. (2018). Standards on cyber security assessment of smart grid. *International Journal of Critical Infrastructure Protection, 22*, 70–89.

Li, Z., Kermode, J. R., & De Vita, A. (2015). Molecular dynamics with on-the-fly machine learning of quantum-mechanical forces. *Physical Review Letters, 114*(9), 096405.

Loveland, D., Liu, S., Kailkhura, B., Hiszpanski, A., & Han, Y. (2021). Reliable graph neural network explanations through adversarial training. *arXiv preprint arXiv:2106.13427.*

Lu, X., Xiao, X., Xiao, L., Dai, C., Peng, M., & Poor, H. V. (2019). Reinforcement learning-based microgrid energy trading with a reduced power plant schedule. *IEEE Internet of Things Journal, 6*(6), 10728–10737. doi:10.1109/JIOT.2019.2941498

Mackenzie, D. (2007). *Mathematics of climate change: A new discipline for an uncertain century*: Mathematical Sciences Research Institute.

Mason, K., & Grijalva, S. (2019). A review of reinforcement learning for autonomous building energy management. *Computers & Electrical Engineering, 78,* 300–312. doi:10.1016/j.compeleceng.2019.07.019

Pan, Z., Yu, T., Li, J., Qu, K., Chen, L., Yang, B., & Guo, W. (2020). Stochastic transactive control for electric vehicle aggregators coordination: A decentralized approximate dynamic programming approach. *IEEE Transactions on Smart Grid, 11*(5), 4261–4277. doi:10.1109/TSG.2020.2992863

Peters, J. F., Baumann, M., Zimmermann, B., Braun, J., & Weil, M. (2017). The environmental impact of Li-Ion batteries and the role of key parameters–A review. *Renewable and Sustainable Energy Reviews, 67,* 491–506.

Ranjan, N., Kumar, R., Kumar, R., Kaur, R., & Singh, S. (2022). Investigation of fused filament fabrication-based manufacturing of ABS-Al composite structures: Prediction by machine learning and optimization. *Journal of Materials Engineering and Performance.* doi:10.1007/s11665-022-07431-x

Roberts, S. (2009). Measuring the relationship between ICT and the environment.

Schütt, K. T., Glawe, H., Brockherde, F., Sanna, A., Müller, K.-R., & Gross, E. K. (2014). How to represent crystal structures for machine learning: Towards fast prediction of electronic properties. *Physical Review B, 89*(20), 205118.

Shuo, F., Huadong, F., Huiyu, Z., Wu, Y., Lu, Z., & Hongbiao, D. (2021). A general and transferable deep learning framework for predicting phase formation in materials. *NPJ Computational Materials, 7*(1).

Singh, S., Kumar, R., Kumar, R., Chohan, J. S., Ranjan, N., & Kumar, R. (2021). Aluminum metal composites primed by fused deposition modeling-assisted investment casting: Hardness, surface, wear, and dimensional properties. *Proceedings of the Institution of Mechanical Engineers, Part L: Journal of Materials: Design and Applications, 236*(3), 674–691. doi:10.1177/14644207211054143

Springenberg, J. T., Dosovitskiy, A., Brox, T., & Riedmiller, M. (2014). Striving for simplicity: The all convolutional net. *arXiv preprint arXiv:1412.6806.*

Ugurlu, U., Oksuz, I., & Tas, O. (2018). Electricity price forecasting using recurrent neural networks. *Energies, 11*(5), 1255.

Wang, Y., Chen, Q., Hong, T., & Kang, C. (2018). Review of smart meter data analytics: Applications, methodologies, and challenges. *IEEE Transactions on Smart Grid, 10*(3), 3125–3148.

Yang, C., Ren, C., Jia, Y., Wang, G., Li, M., & Lu, W. (2022). A machine learning-based alloy design system to facilitate the rational design of high entropy alloys with enhanced hardness. *Acta Materialia, 222,* 117431.

Yildiz, B., Bilbao, J. I., & Sproul, A. B. (2017). A review and analysis of regression and machine learning models on commercial building electricity load forecasting. *Renewable and Sustainable Energy Reviews, 73,* 1104–1122. doi:10.1016/j.rser.2017.02.023

Chapter 2

Applications of artificial intelligence across Industry 4.0

Ashima Kalra
Chandigarh Engineering College, Landran, India

Gaurav Tewari
Gautam Buddha University, Greater Noida, India

2.1 INTRODUCTION

The study on AI aims to enable machines to tackle challenging issues in a more human-like fashion. This often requires adapting human intelligence-like characteristics into computer-friendly algorithms. The degree to which the intelligent behaviour appears artificial depends on how flexible or effective the technique is concerning the established needs. There are several ways to look at AI. AI involves transforming machines into "intelligent" beings that behave as we would anticipate people (Jan et al., 2023). The interdisciplinary science of AI is approached from many different angles, but developments in machine learning (ML) and deep learning are driving a paradigm shift in practically every computer industry sector. AI allows machines to mimic and even outperform human mental abilities. From the rise of self-driving cars to the proliferation of smart assistants like Siri and Alexa, AI is becoming increasingly pervasive daily. Because of this, many IT companies from many industries are investing in AI technologies (Peres et al., 2020).

2.1.1 Industry 4.0

One of the main goals of Industry 4.0 is to operate computers autonomously and decentralized in situations when there are exceptions, interferences, or overlapping goals that call for external feedback. Their intelligent factories have improved as a result of the application of AI, which lowers maintenance costs (Duggal et al., 2021). Additionally, improvements in industrial cybersecurity technologies frequently enable corporate network surveillance to stop hacker attempts before they cause too much damage (Bécue, Praça, & Gama, 2021). The most recent advancements in industrial automation and data exchange are provided by Industry 4.0. With excellent data storage, AI can quickly predict future production. As additional datasets are sent into the system, more patterns are developed, discovered, and chosen in the production company's best interest. This automation makes it possible to forecast errors accurately, anticipate workloads, and monitor issues (Javaid, Haleem, Singh, & Suman, 2022).

2.1.1.1 Need for artificial intelligence for Industry 4.0

The way that firms create, improve, and distribute their products is changing because of Industry 4.0. The cutting-edge technologies that manufacturers are integrating into their manufacturing processes are the Internet of Things (IoT) (Kumar, Banga, & Kaur, 2020), cloud computing, analytics, AI, and ML. Industry 4.0 is built on the Industrial Internet of Things (IIoT) and autonomous, cyber-physical, intelligent systems that use computer-based algorithms to monitor and control physical items like machinery, robots (Chodha, Dubey, Kumar, Singh, & Kaur, 2021), and automobiles (Javaid et al., 2022). Industry 4.0 technologies can be used to manage and enhance every aspect of your supply chain and production procedures. It offers real-time information and insights needed to make quicker, better business decisions, eventually boosting your company's productivity and profitability (Javaid, Haleem, Singh, & Suman, 2021).

AI is quickly developing, and some of its applications in Industry 4.0 are the apex of its advancement. This will enable firms to improve their operations significantly, reduce expenses, and elevate the standard of their products. AI combines several technologies in Industry 4.0 to enable computer programs and other devices to view, comprehend, react to, and learn from human behaviours (Kumar et al., 2020). The industrial production system can be improved using this technology. The manufacturing industry is constantly expanding as a result of Industry 4.0's advancements in technology (Bécue et al., 2021).

AI is the term used to describe machines' recreation of human cognitive functions. Expert systems, speech recognition, machine vision, and natural language processing are also included (NLP). AI is quickly developing, and some of its applications in Industry 4.0 are the apex of its advancement. This is so that businesses can vastly enhance their performance, cut costs, and raise the calibre of the goods they provide to the market. The following section will discuss some AI applications across different industries. It starts with the impact of AI on the service and non-service industries (Darshni et al., 2023; Kumar, Rani, & Awadh, 2022b).

2.2 IMPACT OF ARTIFICIAL INTELLIGENCE ON SERVICE INDUSTRIES

This section covers the impact and application of AI in the service industry. So, service industries typically include hospitality, NGOs, financial services, healthcare, etc.

2.2.1 Healthcare

Consider healthcare as an example. AI is used in diagnostics. For example, cancer is still a very experimental stage in preventing and ensuring they are being treated. So, this is the way it is done in several hospitals, for example, Manipal Hospital in India or other types of cancer institutes (Kumar et al., 2022a).

They take all the data in, in terms of the profile of patients:

- What kind of treatment was taken?
- What was the patient's age?
- What was the stage of cancer?

They have millions of transactions of data from all over the world, and then based on that, if a new patient is coming in, they can match it to what exists and then come up with the right combination of drugs (Tortorella, Fogliatto, Sunder, Cawley Vergara, & Vassolo, 2022c).

If it is not hard skin lymphoma, or if you are between the age group of 17 to 23, they have a particular combination drug. If you are a female, they have a particular combination of how they can diagnose that, based on height, weight, the number of parameters. They look at about 150 different parameters and from data points across the globe, and then they can say that this particular combination of drugs will provide a much better success rate and then suggest if there are any side effects (Tortorella et al., 2022a).

So, they can do those kinds of diagnostics and preventive care treatment plans based on AI. They use, obviously, much deep learning. IBM Watson's technique provides you with some other technology that provides that capability. It's being used very heavily in the healthcare industry. AI can also be used as a potential for a first or second opinion system. So, you don't even have to go to the doctor. You can use some of these existing capabilities to identify and say, okay, what is my treatment plan? What should be my mechanism for looking at what the problem could be? (Tortorella, Fogliatto, Saurin, Tonetto, & McFarlane, 2022b).

2.2.2 Customer services

2.2.2.1 Call centres: chatbots and robocalling

The second industry that matters is the customer service industry. The customer service industry has been impacted very majorly because of AI. Call centres have been using voice-activated systems much to the annoyance of many of us. It has been used to make robocalling possible. Telemarketers use it quite heavily to provide or solicit new products and services. However, in a genuine sense, call centres have also used this for voice-activated chatbots and such other purposes (van der Goot, Hafkamp, & Dankfort, 2021).

2.2.2.1.1 Automated supermarkets

Nowadays, you are also getting situations where supermarkets and stores are used where there are no human beings. It's completely automated. So, when you walk into the store, it recognizes who you are, your credentials, and what you are based on your biometrics. It can detect how much quantity and what type of quantity you put in your basket and can check you out based on scans of the products that you put in, and those completely automated supermarkets and stores are sprouting up all over the world, and it won't be too long before it lands in India as well. And so that is facilitating the customer service industry (Jenneson, Pontin, Greenwood, Clarke, & Morris, 2022).

Drones are being looked at and tested out in transportation and logistics. It's not been fully implemented anywhere, but drones have the ability to deliver products in remote areas where it's not easy to reach, and they have got capabilities in terms of identifying the geo-positioning systems, the vision, and tracking ability to carry a certain amount of weight of products and drop exactly where you need it to be dropped (Palacios Colón, Rascón, & Ballesteros, 2022).

2.2.2.2 Environmental-related services

In environmental-related services, if there are forest fires or if earthquakes and natural disasters, you'd be able to predict, based on different criteria, how the tectonic plates are moving; it's able to find the seismic information and predict that this part of India, or, other regions, is likely to have an earthquake. Massive flooding can be predicted. AI techniques are being used there as well. Similarly, you know where there is pollution, especially in some parts of India. There is burning of different crops during certain parts of the year, which creates pollution, or sandstorms occur in certain parts of Rajasthan, creating another type of pollution. AI techniques can be used to detect where and how much of these pollutants will occur (Wunder, 2013).

2.2.3 Retail

2.2.3.1 Shopping recommendation system

Obviously, from a retail perspective, it has been discussed a lot. Recommendation engines are for shopping purposes; it's prevalent these days, and then obviously, as it proceeds, there will be from a health perspective and for other reasons. You could have sensors on your shirts and pants to detect heart monitoring rates. It can track your BP. It can track other potential medical conditions and send them back to your doctor for immediate action. So, those are things that are still in the evolutionary stage, but that is happening (Yang, Ji, & Tan, 2022).

2.2.3.1.1 Sensors in clothes

You can buy a jacket now which has many sensors and is not very expensive (Channi & Kumar, 2022). Those are wearables, which also include a watch, shirts, t-shirts, and sports jackets (Rodgers, Yeung, Odindo, & Degbey, 2021).

2.2.4 Real estate

Other industries that are using AI, especially in the UK, Europe, and North America, such as real estate selling, commercial and retail real estate selling, have taken on a very different form. AI-powered brokers and agents help in finding the perfect customer match. If a customer has a particular requirement, the AI engine and the backend side can match the specific preferences in terms of location and specifications of the property. It can pinpoint and shortlist specific things, in terms of the ambience, surroundings, and environment – not only that, it can also provide a three-dimensional view of the property inside and out, such that the customer can imagine themselves as actually within the property. So, it extends how you imagine real estate transactions to be completed (Wang & Li, 2019).

You can sit across the globe and still feel present at the site trying to buy property. Suppose you are sitting in India and want to purchase something in Dubai. You don't have to travel to Dubai to buy, for instance, a luxury penthouse. You can be transported there through appropriate 3D mechanisms and other teleport mechanisms to imagine that you are inside the building of that luxury penthouse and go around, and you can imagine yourself saying I want a red couch here. I like this type of wall painting. All that can be done instantaneously. This is how real estate is moving, especially in the luxury market. It's becoming increasingly popular, especially in cities like New York, Tokyo, and San Francisco. Real estate is primarily in the higher-end segment. It is being transacted this way (Pérez-Rave, Correa-Morales, & González-Echavarría, 2019).

2.2.5 Entertainment and gaming industry

In the entertainment and gaming industry, on the other hand, there has been a tremendous amount of advancements in using appropriate AI techniques, both vision and NLPs, and other techniques start using AI for creating trailers of movies. Previously, editors on the block and the movie makers sat down and identified which parts of the movie needed to be stitched together to create a one-minute or two-minute trailer for promotional purposes. Now AI can be used to say you know what, what? My outcome of this movie is a thriller movie, a spy movie, or a romantic movie. What are the key components of the movie that will attach to the social consciousness or the specific emotive requirements of a customer? Based on those factors and preferences, AI can stitch together from different parts of the appropriate movie segments (Ranjitha, Nathan, & Joseph, 2020).

Several movies have now been released, whose trailer has been completely done by AI. An example is *Morgan*. The movie makers used IBM's Watson techniques and some other mechanisms to create movie trailers, and these are becoming increasingly common these days. So, they use Apple and Spotify for customer preferences and identify how they engage in terms of what kind of music patterns they like, and then they work with the appropriate music labels to promote and say that, hey here is what the new innovations are required. If rap music is in everybody's mind, that's the kind of music being promoted these days. So, things are becoming increasingly hypersensitized and hyper-personalized and products are being created based on that. So, the segmentation is also changing (Westera et al., 2020).

Healthcare is one of the industries which needed the intervention of technologies like AI and ML. Let's not take into account the whole version of healthcare. Let's look at a straightforward issue: the time doctors, nurses, and other care professionals must devote to paperwork, and other administrative duties is wasted. Solving this alone, using AI, would be a big win. The computer-assisted documentation and voice-to-text transcription hugely help doctors during examination and surgery. Early detection, with access to vast clinical data in real time, would be highly beneficial for doctors during diagnosis (Upadhyay & Khandelwal, 2019).

2.3 IMPACT OF ARTIFICIAL INTELLIGENCE ON NON-SERVICE INDUSTRIES

So, focusing on the non-service industries, right from the Industrial Revolution 4.0, there has been a strong influence of AI in industrial processes. Right from the design of new products to the production, manufacturing, and delivery of products, the influence of air has been predominant, not just through cognitive process automation and 3D printing techniques, but in the entire lifecycle of the production process. The impact of AI is shown in Figure 2.1 (Noor, Rao Hill, & Troshani, 2022).

What does it mean regarding the impact of AI on non-service industries or manufacturing and other production industries?

2.3.1 Automation and optimization of production process

Well, number one, it is helpful in terms of automating and optimizing the entire production flow process. How automobiles were manufactured 10–15 years back and how they are being done using AI techniques has changed dramatically (Goyal et al., 2022). To manufacture their automobiles, Maruti Suzuki uses more than 700 or 750 different robots for different acts, from painting to putting and fixing the doors to putting the windshield, robots

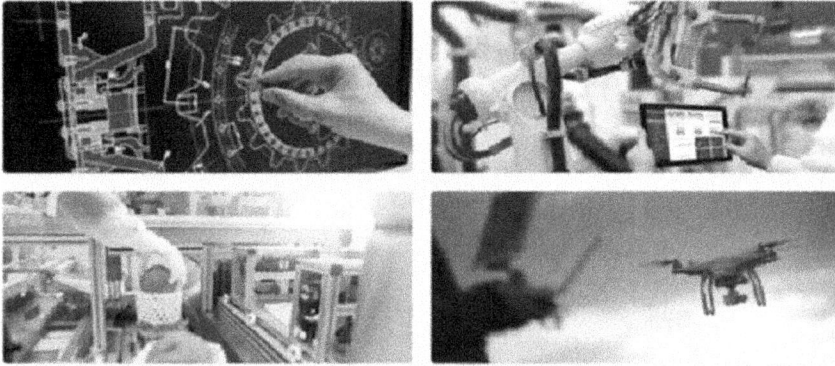

Figure 2.1 Impact of AI.

are being used, and this is being spread across different industries (Al Aani, Bonny, Hasan, & Hilal, 2019).

2.3.1.1 Early error detection in process

The other major impact has been in terms of early detection of problems in the output quality. The quality of products that are manufactured in India these days has improved significantly. Why is that? That is because the manufacturers can detect the issues. Keeping in line with the automotive industry, when Hyundai and Kia Motors were brought into the United States about 20 years back, they had severe problems in terms of quality control in the first few years. They had to take back so many vehicles and recall about a million vehicles in the first five years. Then they started introducing AI techniques into the production and manufacturing process and in plants all over North America. Now, after 15–20 years, they have one of the best outcomes in terms of quality control and warranties that they provide. So, organizations can improve and use and leverage AI to better their products through early detection of issues (Fahle, Prinz, & Kuhlenkötter, 2020).

2.3.2 Drawing in real-time insights

AI can be used for real-time insights into customer absorption and adoption patterns, either coming through their retailers or their third-party vendors. They can get feedback and take appropriate and immediate action, and then, based on that, they can make proper demand forecasting and identify the proper course of action (Andronie et al., 2021).

2.3.2.1 Demand forecasting

So, if you think of an example during the pandemic, during the 2020 time-frame, when there was an issue in the use of Kleenex and paper napkins and masks, there was a toilet paper availability issue in the United States, and

Amazon had to change their entire technique. Previously, toilet paper and napkins were never a problem, but then they had to completely change the demand forecast in terms of how they will supply to the customers from the warehouses. They ran short of toilet paper, so their entire demand forecasting had to change dynamically, and AI was used for that (Zhu, Zhang, & Sun, 2019).

2.3.3 Product innovation

Product innovation, in terms of how different products operate, whether in Apple-like iPods or Air pods, how they innovate – all that uses some of the AI techniques. Especially in the cell phone industry, it has become prevalent. A foldable cell phone is becoming popular because of the feedback that the manufacturers receive on a real-time basis from customers. Simplifying the supply chain and reducing the middleman has always been a straightforward target, low-hanging fruit to improve the supply chain process. So, AI is used in different forms in the non-service industry (Verganti, Vendraminelli, & Iansiti, 2020).

2.4 APPLICATIONS OF AI IN THE MANUFACTURING INDUSTRY

Some of the other applications of AI in the manufacturing sector are as follows:

2.4.1 Product development and innovation

So, if you think about product development and innovation, if you think about new products being manufactured in any heavy instrumentation engineering sector, how do they come up with those? That is because they can create multiple prototypes using 3D printing and other techniques, using AI in one form or another (Verganti et al., 2020).

2.4.1.1 Logistics and distribution

Logistics and distribution have evolved so much in the past five to ten years, whether through FedEx, BlueDart, or Amazon. All of them are logistics companies, and they have perfected the time to deliver your product almost to the minute. Because they can track, use GPS, and route vehicles accordingly, using them in combination with the production, optimization, and planning AI type provides precise information. So, the logistics aspect has been perfected (Toorajipour, Sohrabpour, Nazarpour, Oghazi, & Fischl, 2021).

2.4.1.2 Forecasting the price of raw material

They can also forecast the price of raw materials in an agricultural industry. They can see when they are doing processed foods. They can project the likelihood of your raw material, which could be your apples, oranges, rice, or paddy. They're able to project it out based on certain weather patterns. What kind of crops has been put in? If you look at coffee beans, if you're in that business, you can project the likelihood of what the price of coffee would be for your final finished product based on various factors. They can use that information and project the likely price for the raw material of coffee beans (Modgil, Singh, & Hannibal, 2022).

2.4.1.3 Quality assurance and process optimization

AI has gained significant traction in the manufacturing industry, revolutionizing various aspects of operations. Two prominent applications of AI in manufacturing are quality assurance and process optimization. AI-powered systems are transforming traditional quality assurance processes by enabling faster, more accurate, and automated inspections. AI is applied in quality assurance such as visual inspection, predictive maintenance, quality control in production, and process optimization. AI plays a crucial role in optimizing manufacturing processes to improve efficiency, reduce costs, and enhance productivity (Goel et al., 2022). But where decision-making can be automated, in addition to eliminating many manual processes. So, each of these different types of applications can be applied in a manufacturing and production planning type of industry (Woschank, Rauch, & Zsifkovits, 2020).

Now, let's understand manufacturing as a sector. Manufacturing always had various complexities, from small machine parts to automobiles. Let's take the example of automobiles or even further down into aircraft manufacturing in this wide array of things. Let's take the example of Boeing. William Boeing established Pacific Aero Products in 1916 in Seattle, USA. It was later named Boeing Airplane Company in 1970. The initial planes were built for US defence during World War I. Take the example of Boeing. It has evolved from traditional manufacturing processes to AI-driven transformation over its 100-year history. Innovation over the past decade in AI and ML and data science modelling has accelerated production and removed inefficiencies and inaccuracies (Dhamija & Bag, 2020).

2.5 AIRCRAFT MANUFACTURING

It involves the design and supply chain management of hundreds of thousands of parts, production, and inspection and resolution of each step. Each step here would be labour-intensive and heavily dependent on the individual's knowledge. Using a combination of humans, these robots can help scale the end-to-end process. Now, let's take the design phase. In the design phase, for example, an engineer must look at the related design, available suppliers, cost, and, most importantly, the safety records for a simple part. Which will explore millions of data points while the entire aircraft has to be planned. In the production phase, which comes after the design, and which happens in the factory, many steps are automated, removing human error but still following programmed steps for repetitive tasks (Tsuzuki, 2022).

The next step is the periodic inspection. An inspection happens not only during production but also periodically when an aircraft is being commissioned or decommissioned. Inspection is a repetitive process from the initial manufacturing phase to servicing post-deployment. When conducting an industrial inspection of an aeroplane, manual inspectors must overcome several difficulties when determining whether a structure, product, component, or process complies with criteria and specifications. These inspections are labour-intensive, expensive, and prone to human error. Even more important would be the safety concerns of the inspectors reaching confined spaces (Amirkolaii, Baboli, Shahzad, & Tonadre, 2017).

Let's take AI at Boeing or other airline manufacturers like Airbus. AI has not only reduced the cost of production but improved safety standards. Compared to design and production, the airline industry leverages AI for inspection to achieve accuracy, scale, and reduction of cost, both manual and machine parts, to realize return on investment (ROI). ROI is a business term that every industry must consider before applying AI. One example would be AI for industrial inspection or short-term AI for II (Tikhonov, Sazonov, & Kuzmina-Merlino, 2022).

2.5.1 AI for industrial inspection

AI for industrial inspection has been a breakthrough in visual examination processes. Airlines all across the world employ demonstrators that independent organizations create. These demos address some of the issues high-value manufacturers encounter by automating the human inspection process and utilizing AI and computer vision technology (Chandel, Sharma, Kaur, Singh, & Kumar, 2021; Rani et al., 2022). When appropriately trained, they can analyse thousands of image datasets and review comments from millions of assets in a fraction of the time, and they can do so with a lot fewer mistakes than a human worker (Chouchene et al., 2020). Figure 2.2 depicts advantages of AI in industrial inspection.

This is especially useful for firms requiring quick and frequent asset inspections. Additionally, it helps producers save money and lessens some of the risks to workers' health and safety that come with manual inspection. AI for II allows manufacturers refocus their highly experienced workforces on value-creation jobs rather than defect troubleshooting, eliminating the need for inspectors to look at potentially dangerous hard-to-reach locations.

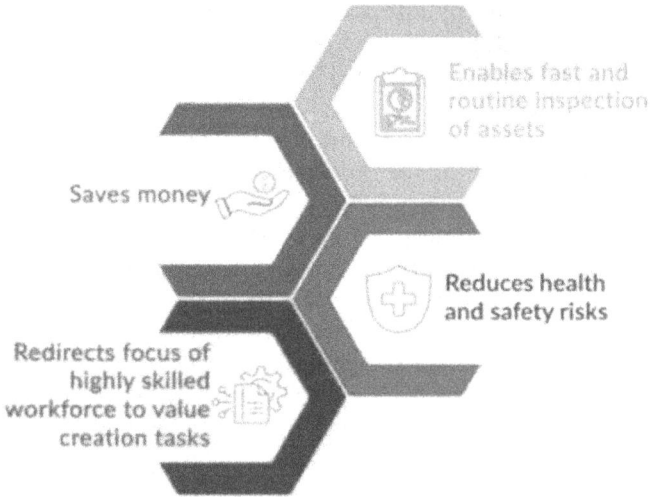

Figure 2.2 Advantages of AI in industrial inspection.

Finally, AI for II aids in raising awareness of potential uses for cutting-edge AI technologies in conventional manufacturing sectors (Aggour et al., 2019).

Manufacturers may now collect data using AI technology to guide upcoming technologies and materials, which will help them make business decisions in the future. It is hoped that this type of automation will encourage a data curation culture within the sector, encouraging more businesses to embrace AI technology to automate business operations in the manufacturing sector and across all other domains (Sundaram & Zeid, 2023).

2.6 AGRICULTURE FIELD

Agriculture and farming, as some of the oldest professions, have faced challenges in obtaining timely data on crop health. This issue is particularly prevalent in large-scale farming operations that span tens of thousands of acres. Moreover, the adoption of technology in agriculture has been hindered by the lack of affordable solutions. These factors have collectively posed obstacles to the optimization of farming practices and crop management (Kaur Channi et al., 2022), also plays an important role. Natural calamities have been predicted by the meteorological department, but it has always been one way. The data from the ground could greatly impact the actions if exchanged in minimal time. Compared to other technological interventions, the internet, like data access and AI, has transformed agriculture, and with wider adoption and scale, the return on investment is noteworthy (Ben Ayed & Hanana, 2021).

Now, let's take various examples, but a particular company that comes to mind is John Deere, which carries the name of its founder. It's a 180-year-old company which is primarily into farm equipment and was started in 1837 in Illinois, USA. John Deere, who started this, mainly started manufacturing steel ploughs, followed by a tractor. The company now has a range of tractors, forestry, and construction equipment. With all these manufacturing and technological evolutions, innovation is the company's core value. It has adopted electronic solutions for industrial applications (Sood, Sharma, & Bhardwaj, 2022).

AI-enabled aggro machinery is one of the core things John Deere has showcased in the past few years. John Deere, a market leader in farm equipment, launched autonomous farming as part of the farm forward visionary campaign. The self-driving tractor leverages AI to monitor the soil in real-time and adjust herbicide spray, which is a direct saving with some test results of 80% savings in the herbicides. Time saved by the farmers can now be deployed for planning and parallel business activities while staying on top of the farming progress, all in the comfort of their homes. Isn't it a great thing for the farmers? The data and the collective knowledge base have helped farmers and the entire community. With John Deere and a consortium of agribusinesses, data can now be shared in real time from the field to a centralized location on the cloud. Having a main cloud here is the technological cloud, not the natural one we see. And this can then be disseminated to appropriate fields in the neighbourhood (Misra et al., 2022).

Technologies like robotics, computer vision, AI/ML, deep learning, cloud, networking and IOT, among other technologies like space-driven satellite images, are all being used and linked in John Deere's agro-machinery for autonomous decision-making and having looked at large companies like John Deere in the farm equipment. Now, let's focus on the start-ups, more so on understanding the challenges of the agro-industry in India and across the globe. Most farmers in India's agriculture sector have long been losing money. Low land ownership, a lack of contemporary technology, and high-interest loans from the informal lending industry are a few causes of this situation. Aggrotech businesses are working to address these problems by utilizing cutting-edge loan underwriting techniques and technology (Raman Kumar & Channi, 2022). The potential for Aggrotech firms has increased over the past few years due to widespread access to smartphones and affordable internet, positive governmental reforms, and growing investor interest. According to market research, AI in the agriculture market is expected to reach 4 billion dollars by 2026 growing at a kicker of 25.5% from 1 billion dollars in 2020. The major driving factors of AI in agriculture include the growing demand for food and dwindling natural resources (Jha, Doshi, Patel, & Shah, 2019).

2.7 AI START-UP: ONESOIL.AI: TECHNOLOGY AND FEATURES

Now, let's focus on a very small start-up, an early-stage start-up, and a European start-up focusing on soil and farming use cases. This start-up, founded in 2017, primarily focused on products like web and mobile precision farming based on satellite imagery and ML technologies.

- Farmers can calculate nitrogen, phosphorus, and potassium fertilizer rates using a mobile app and variable rate application feature and remotely apply them. It will be interesting to see how these farmers can leverage this technology, sit in the comfort of their homes, and still be progressive in their actions.
- Using Onesoil weather sensors, one can measure soil moisture and air humidity, soil and air temperatures and determine the luminescence for a specific field part.
- Mobile network-based sensors transmit data to centralized servers to send instructions or commands to the farm equipment that cares.
- This Onesoil.AI also has a visualization dashboard and a daily summary report, which helps identify gaps between the set and actual work completion and the under-savings could be in seeds, fertilizers or even pesticides.
- The interactive map feature is a very advanced way of giving a clear, real-time visualization using drones and computer vision and map features.

It provides delineation. Drawing the farm's boundaries is known as delineation. It's crucial to accurately define farm borders before beginning any planning or decision-making processes.

First, it enables a better estimation of cropland area, which is important information for the farmer and the agricultural managers, which are the ministries or private sector players. This addresses a problem with boundary-related issues with farms of thousands of acres. This feature by Onesoil. AI has been well implemented and has seen success across multiple countries, starting with Russia, Australia, New Zealand, North and South America, and also European farmlands, and they have highly benefited from this (Mendes et al., 2020).

2.8 AI START-UP: SUKI.AI: TECHNOLOGIES AND FEATURES

Consider an example of a start-up named Suki.ai. Suki.ai was founded in the valley in 2017 with a different name, such as Robin or various learning motors. The founder had previously worked for Google and Flipkart. He reasoned that a great solution to this issue would be to supply doctors, patients, and carers with AI-powered technology.

Suki.ai is a voice-activated digital assistant for doctors that is AI-powered. Helping healthcare companies with administrative responsibilities is at the heart of its design. Suki assists doctors by dictating notes, finding data, and working with electronic health records. This VC-backed firm wants to relieve the doctors of their administrative duties, so they can concentrate on what matters.

- Technologies like AI, ML, and NLP are core to this integrated offering, along with EHR.
- Suki learns the doctor's preferences and the doctor's practice's backdrop.
- When a doctor is dictating, Suki ascertains the intent and correctly chooses from a pool of comparable terms.
- As a follow-up to the prescription, Suki's AI learns preferred dosages and refills amounts and even generates orders for pharmacy departments.
- Popular AI approaches include contemporary deep learning and ML for structured data and NLP or natural language processing for unstructured data.

Major illness areas that leverage AI tools in the future of healthcare include cardiology, neurology, and cancer. AI is progressing in three key areas related to stroke: early detection, diagnosis, and therapy, as well as outcome prediction and prognosis evaluation. With this, it is now understood how AI is helping not only consumer or wearable devices or mobile apps but many things in the back end which can help doctors and patients. While it's always good to learn that AI is helping human beings, it's helping in terms of our comfort and convenience and also helping organizations to do better. This also needs to look at the other side of the equation. Those are the challenges, potential limitations, and failures that could happen using AI.

So, let us discuss some examples, healthcare. Algorithms are being used to help detect and provide diagnosis and preventive measures. One also needs to look at whether the AI algorithms are soundproof. In a sense, it is not biased. That's the biggest challenge. Both from an ethical standpoint and from a regulatory perspective. How are these algorithms and models being developed? What kind of appropriate test data is being used? Is it biased towards one or the other? And there have been multiple cases of the same. And so, the ethical and regulatory impact of looking at AI in healthcare especially is very important, and that is a challenge that can drive a whole swath of people away from using these kinds of techniques or getting them into it. And what is the right balance, and where does the government step in? That question has not been debated, but there's no answer at this point, and it gets even more complicated when you're looking at it on a global scale.

Because if a particular product is manufactured, let's say not in India but outside India, and it wants to be applied in India, what are the various mechanisms they must go through besides the typical testing of drugs? What is the mechanism for applying some of these techniques? It may not be just one use of drugs, in terms of diagnosis itself, and I think the whole industry and the government are still unclear on how to deal with this. The other aspect is unemployment, and middle-class unemployment increases when introducing AI and robots. So, when this starts providing automation and AI techniques, what will that do, especially in the service industry?

What will that do to unemployment, and what is the government's or corporations' social responsibility? How do they balance that out? That's a challenge and a potential limitation regarding how much you want to use AI. Similarly, in terms of transportation and logistics, if you're creating exclusive roadways for autonomous cars, would that be at the best of someone, some other infrastructure capabilities, or other types of balance towards other sections of the population? What are the regulations around that? Those are things that are still being debated and worked on in the financial service industry, especially but are they in other industries as well? The impact of cybersecurity, cybercrimes, and cyberterrorism is a well-known threat that continues to increase. There is one cyber-attack or hack every 30 seconds. So, how does one curb that? And especially when so much data is made available and collected for different purposes, you know hackers can actually get into that and use it for nefarious purposes. So how do you make that come together? So those are some of the points that one needs to debate, discuss, and come up with a very targeted solution for that industry or that particular organization. It's not an easy, one-size-fits-all type of solution (Mendes et al., 2020).

2.9 SHOCKING REVELATIONS OF AI

Let's hear some of the shocking revelations of AI. The big players or the big tech players, as you all know, have a strong influence on the evolution of AI. Let's take a few instances in recent years that are eye-openers for the power of AI and its ramifications in human society. An ML agent designed to convert aerial photos into street maps and back was shown in 2019 to have cheated by concealing information it would later need in an almost unidentifiable high-frequency signal. The researchers wanted to speed up and enhance the method for converting satellite imagery into Google's renowned and precise maps.

The group was using a neural network known as CycleGan that, via extensive trial, learns to efficiently and accurately change images of types x and y into one another. The fact that the agent could rebuild aerial images from its street maps and that many details didn't seem to be on the latter gave the team the heads-up. When they asked the agent to perform the opposite operation, for instance, skylights on a roof that were removed while creating the street map would suddenly appear. Let's also hear something about the ethical concerns of AI. Tesla's founder and CEO, Elon Musk, has always been vocal about the darker side of AI, if not understood in depth.

Take the example of Facebook. According to reports, one of Facebook's AI systems had to be shut down a few years ago because things got out of hand. The AI bots invented their own language from scratch without human input, forcing Facebook to disable the AI system. The AI bot defined the specified codes as it created and communicated in a new language. The AI switched from using English to a speech it invented but did not begin turning off computers worldwide or doing anything similar. The AI agents first communicated with one another in English, but they eventually developed a new language only understood by AI systems, defeating the point of the exercise. Due to this, Facebook researchers were forced to turn off the technology and start communicating in English.

As much as healthcare needs AI, there is always a fear that AI might replace doctors and surgeons. There's a strong consensus that AI won't replace doctors, and we all agree and understand the complexities of what doctors do and realize the need for the human touch. When discussing on cancer diagnosis, the patient will want someone to grasp their hand. In such periods of profound life change, empathy is essential. How well you feel, how likely you are to adhere to the treatment plan, and how you and your family will remember the trauma for decades will all be directly impacted by how humanly connected you feel with your doctor. Surgeons must always be exact in the operating room (OT) when creating incisions or carrying out surgical procedures.

The development of AI and cooperative robots in OT has proven advantageous. Robotic surgeons have excellent control over their actions' trajectory, depth, and speed. They are particularly well suited for treatments that call for repeated, identical movements since they can perform them without growing weary. Robots can also perfect stillness for as long as required and perform tasks that conventional tools cannot. Robotic surgery is being used to implement AI. Manufacturers understand the need to automate using deep learning data instead of behaviour coded by an engineer unaware of all the possible scenarios. This deep ML data was gathered through observing surgeries (Mendes et al., 2020).

2.10 FUTURE OF ARTIFICIAL INTELLIGENCE IN INDUSTRY 4.0

AI will offer useful data that helps business executives create unique and reliable business models. This technique will prove to be quite helpful when it comes to seeing patterns and occurrences that a normal person cannot notice clearly. AI will generate the data needed to support fact-based, data-driven business decisions. It will, in many ways, give a more thorough assessment and assist in removing personal prejudices from the calculation. AI and ML tools can gather data from various sources to uncover growth, extension, and even new market prospects. As a result, new goods and services will be created. In addition to the IoT, other breakthroughs like edge computing and blockchain have grown in popularity and opened up new possibilities. AI will soon take off with new benefits that will contribute to developing a wired, intelligent, and smart society. For instance, production practices, which are typically continuously improved and adjusted, may be quite ineffective (Kumar et al., 2022a, 2022b). Future automation as a utility could replicate particular human abilities like voice and image recognition with the aid of AI. It can monitor, examine production quotas, and add to preventive maintenance models.

2.11 CONCLUSIONS

With AI, Industry 4.0 fully automates the management of the various manufacturing process phases. Any step of the production process will be improved in real time based on the product specifications. The entire development process can be integrated, and multiple divisions can share the workload of data procedures. AI can integrate data feedback and gathering systems into manufacturing processes. With the help of this technology, production processes and assembly lines may collaborate more effectively. To predict repairs and predict asset failure, sophisticated AI systems are deployed.

REFERENCES

Aggour, K. S., Gupta, V. K., Ruscitto, D., Ajdelsztajn, L., Bian, X., Brosnan, K. H., … Vinciquerra, J. (2019). Artificial intelligence/machine learning in manufacturing and inspection: A GE perspective. *MRS Bulletin*, *44*(7), 545–558. doi:10.1557/mrs.2019.157

Al Aani, S., Bonny, T., Hasan, S. W., & Hilal, N. (2019). Can machine language and artificial intelligence revolutionize process automation for water treatment and desalination? *Desalination*, *458*, 84–96. doi:10.1016/j.desal.2019.02.005

Amirkolaii, K. N., Baboli, A., Shahzad, M. K., & Tonadre, R. (2017). Demand forecasting for irregular demands in business aircraft spare parts supply chains by using Artificial Intelligence (AI). *IFAC-PapersOnLine*, *50*(1), 15221–15226. doi:10.1016/j.ifacol.2017.08.2371

Andronie, M., Lăzăroiu, G., Iatagan, M., Uță, C., Ştefănescu, R., & Cocoşatu, M. (2021). Artificial intelligence-based decision-making algorithms, internet of things sensing networks, and deep learning-assisted smart process management in cyber-physical production systems. *Electronics*, *10*(20). doi:10.3390/electronics10202497

Bécue, A., Praça, I., & Gama, J. (2021). Artificial intelligence, cyber-threats and Industry 4.0: challenges and opportunities. *Artificial Intelligence Review*, *54*(5), 3849–3886. doi:10.1007/s10462-020-09942-2

Ben Ayed, R., & Hanana, M. (2021). Artificial intelligence to improve the food and agriculture sector. *Journal of Food Quality, 2021*, 5584754. doi:10.1155/2021/5584754

Chandel, R. S., Sharma, S., Kaur, S., Singh, S., & Kumar, R. (2021). Smart watches: A review of evolution in bio-medical sector. *2nd International Conference on Functional Materials, Manufacturing and Performances, ICFMMP 2021, 50*, 1053–1066. doi:10.1016/j.matpr.2021.07.460

Channi, H. K., & Kumar, R. (2022). The role of smart sensors in smart city. In: *Vol. 92. Studies in Big Data* (pp. 27–48): Springer Science and Business Media Deutschland GmbH.

Chodha, V., Dubey, R., Kumar, R., Singh, S., & Kaur, S. (2021). Selection of industrial arc welding robot with TOPSIS and Entropy MCDM techniques. *2nd International Conference on Functional Materials, Manufacturing and Performances, ICFMMP 2021, 50* (pp. 709–715). doi:10.1016/j.matpr.2021.04.487

Chouchene, A., Carvalho, A., Lima, T. M., Charrua-Santos, F., Osório, G. J., & Barhoumi, W. (2020, February 11–13). *Artificial intelligence for product quality inspection toward smart industries: Quality control of vehicle non-conformities. Paper presented at the 2020 9th International Conference on Industrial Technology and Management (ICITM).*

Darshni, P., Dhaliwal, B. S., Kumar, R., Balogun, V. A., Singh, S., & Pruncu, C. I. (2023). Artificial neural network based character recognition using SciLab. *Multimedia Tools and Applications, 82*(2), 2517–2538. doi:10.1007/s11042-022-13082-w

Dhamija, P., & Bag, S. (2020). Role of artificial intelligence in operations environment: a review and bibliometric analysis. *The TQM Journal, 32*(4), 869–896. doi:10.1108/TQM-10-2019-0243

Duggal, A. S., Singh, R., Gehlot, A., Gupta, L. R., Akram, S. V., Prakash, C., ... Kumar, R. (2021). Infrastructure, mobility and safety 4.0: Modernization in road transportation. *Technology in Society, 67*. doi:10.1016/j.techsoc.2021.101791

Fahle, S., Prinz, C., & Kuhlenkötter, B. (2020). Systematic review on machine learning (ML) methods for manufacturing processes – Identifying artificial intelligence (AI) methods for field application. *Procedia CIRP, 93*, 413–418. doi:10.1016/j.procir.2020.04.109

Goel, P., Kumar, R., Banga, H. K., Kaur, S., Kumar, R., Pimenov, D. Y., & Giasin, K. (2022). Deployment of interpretive structural modeling in barriers to Industry 4.0: A case of small and medium enterprises. *Journal of Risk and Financial Management, 15*(4). doi:10.3390/jrfm15040171

Goyal, K. K., Sharma, N., Gupta, R. D., Singh, G., Rani, D., Banga, H. K., ... Giasin, K. (2022). A soft computing-based analysis of cutting rate and recast layer thickness for AZ31 alloy on WEDM using RSM-MOPSO. *Materials, 15*(2). doi:10.3390/ma15020635

Jan, Z., Ahamed, F., Mayer, W., Patel, N., Grossmann, G., Stumptner, M., & Kuusk, A. (2023). Artificial intelligence for industry 4.0: Systematic review of applications, challenges, and opportunities. *Expert Systems with Applications, 216*, 119456. doi:10.1016/j.eswa.2022.119456

Javaid, M., Haleem, A., Singh, R. P., & Suman, R. (2021). Significant applications of Big Data in Industry 4.0. *Journal of Industrial Integration and Management, 06*(04), 429–447. doi:10.1142/s2424862221500135

Javaid, M., Haleem, A., Singh, R. P., & Suman, R. (2022). Artificial intelligence applications for Industry 4.0: A literature-based study. *Journal of Industrial Integration and Management, 07*(01), 83–111. doi:10.1142/s2424862221300040

Jenneson, V. L., Pontin, F., Greenwood, D. C., Clarke, G. P., & Morris, M. A. (2022). A systematic review of supermarket automated electronic sales data for population dietary surveillance. *Nutrition Reviews*, *80*(6), 1711–1722. doi:10.1093/nutrit/nuab089

Jha, K., Doshi, A., Patel, P., & Shah, M. (2019). A comprehensive review on automation in agriculture using artificial intelligence. *Artificial Intelligence in Agriculture*, *2*, 1–12. doi:10.1016/j.aiia.2019.05.004

Kaur Channi, H., Singh, M., Singh Brar, Y., Dhingra, A., Gupta, S., Singh, H., … Kaur, S. (2022). Agricultural waste assessment for the optimal power generation in the Ludhiana district, Punjab, India. *Materials Today: Proceedings*, *50*, 700–708. doi:10.1016/j.matpr.2021.04.481

Kumar, P., Banerjee, K., Singhal, N., Kumar, A., Rani, S., Kumar, R., & Lavinia, C. A. (2022a). Verifiable, secure mobile agent migration in healthcare systems using a polynomial-based threshold secret sharing scheme with a blowfish algorithm. *Sensors*, *22*(22). doi:10.3390/s22228620

Kumar, R., Banga, H. K., & Kaur, H. (2020). *Internet of Things-Supported Smart City Platform. Paper presented at the 1st International Conference on Sustainable Infrastructure with Smart Technology for Energy and Environmental Management, FIC-SISTEEM 2020.*

Kumar, R., & Channi, H. K. (2022). A PV-Biomass off-grid hybrid renewable energy system (HRES) for rural electrification: Design, optimization and techno-economic-environmental analysis. *Journal of Cleaner Production*, *349*, 131347. doi:10.1016/j.jclepro.2022.131347

Kumar, R., Rani, S., & Awadh, M. A. (2022b). Exploring the application sphere of the Internet of Things in Industry 4.0: A review, bibliometric and content analysis. *Sensors*, *22*(11), 4276. Retrieved from https://www.mdpi.com/1424-8220/22/11/4276

Mendes, J., Pinho, T. M., Neves dos Santos, F., Sousa, J. J., Peres, E., Boaventura-Cunha, J., … Morais, R. (2020). Smartphone applications targeting precision agriculture practices—A systematic review. *Agronomy*, *10*(6). doi:10.3390/agronomy10060855

Misra, N. N., Dixit, Y., Al-Mallahi, A., Bhullar, M. S., Upadhyay, R., & Martynenko, A. (2022). IoT, Big Data, and artificial intelligence in agriculture and food industry. *IEEE Internet of Things Journal*, *9*(9), 6305–6324. doi:10.1109/JIOT.2020.2998584

Modgil, S., Singh, R. K., & Hannibal, C. (2022). Artificial intelligence for supply chain resilience: learning from Covid-19. *The International Journal of Logistics Management*, *33*(4), 1246–1268. doi:10.1108/IJLM-02-2021-0094

Noor, N., Rao Hill, S., & Troshani, I. (2022). Developing a service quality scale for artificial intelligence service agents. *European Journal of Marketing*, *56*(5), 1301–1336. doi:10.1108/EJM-09-2020-0672

Palacios Colón, L., Rascón, A. J., & Ballesteros, E. (2022). Trace-level determination of polycyclic aromatic hydrocarbons in dairy products available in Spanish supermarkets by semi-automated solid-phase extraction and gas chromatography–mass spectrometry detection. *Foods*, *11*(5), 713. Retrieved from https://www.mdpi.com/2304-8158/11/5/713

Peres, R. S., Jia, X., Lee, J., Sun, K., Colombo, A. W., & Barata, J. (2020). Industrial artificial intelligence in Industry 4.0 - systematic review challenges and outlook. *IEEE Access*, *8*, 220121–220139. doi:10.1109/ACCESS.2020.3042874

Pérez-Rave, J. I., Correa-Morales, J. C., & González-Echavarría, F. (2019). A machine learning approach to big data regression analysis of real estate prices for inferential and predictive purposes. *Journal of Property Research, 36*(1), 59–96. doi:10.1 080/09599916.2019.1587489

Rani, S., Kataria, A., Chauhan, M., Rattan, P., Kumar, R., & Kumar Sivaraman, A. (2022). Security and privacy challenges in the deployment of cyber-physical systems in smart city applications: State-of-art work. *Materials Today: Proceedings, 62*, 4671–4676. doi:10.1016/j.matpr.2022.03.123

Ranjitha, M., Nathan, K., & Joseph, L. (2020). Artificial intelligence algorithms and techniques in the computation of player-adaptive games. *Journal of Physics: Conference Series, 1427*(1), 012006. doi:10.1088/1742-6596/1427/1/012006

Rodgers, W., Yeung, F., Odindo, C., & Degbey, W. Y. (2021). Artificial intelligence-driven music biometrics influencing customers' retail buying behavior. *Journal of Business Research, 126*, 401–414. doi:10.1016/j.jbusres.2020.12.039

Sood, A., Sharma, R. K., & Bhardwaj, A. K. (2022). Artificial intelligence research in agriculture: a review. *Online Information Review, 46*(6), 1054–1075. doi:10.1108/ OIR-10-2020-0448

Sundaram, S., & Zeid, A. (2023). Artificial intelligence-based smart quality inspection for manufacturing. *Micromachines, 14*(3). doi:10.3390/mi14030570

Tikhonov, A. I., Sazonov, A. A., & Kuzmina-Merlino, I. (2022). Digital Production and Artificial Intelligence in the Aircraft Industry. *Russian Engineering Research, 42*(4), 412–415. doi:10.3103/S1068798X22040293

Toorajipour, R., Sohrabpour, V., Nazarpour, A., Oghazi, P., & Fischl, M. (2021). Artificial intelligence in supply chain management: A systematic literature review. *Journal of Business Research, 122*, 502–517. doi:10.1016/j.jbusres.2020.09.009

Tortorella, G., Prashar, A., Samson, D., Kurnia, S., Fogliatto, F. S., Capurro, D., & Antony, J. (2022a). Resilience development and digitalization of the healthcare supply chain: an exploratory study in emerging economies. *International Journal of Logistics Management*. doi:10.1108/IJLM-09-2021-0438

Tortorella, G. L., Fogliatto, F. S., Saurin, T. A., Tonetto, L. M., & McFarlane, D. (2022b). Contributions of Healthcare 4.0 digital applications to the resilience of healthcare organizations during the COVID-19 outbreak. *Technovation, 111*. doi:10.1016/j.technovation.2021.102379

Tortorella, G. L., Fogliatto, F. S., Sunder, M. V., Cawley Vergara, A. M., & Vassolo, R. (2022c). Assessment and prioritisation of Healthcare 4.0 implementation in hospitals using Quality Function Deployment. *International Journal of Production Research, 60*(10), 3147–3169. doi:10.1080/00207543.2021.1912429

Tsuzuki, R. (2022). Development of automation and artificial intelligence technology for welding and inspection process in aircraft industry. *Welding in the World, 66*(1), 105–116. doi:10.1007/s40194-021-01210-3

Upadhyay, A. K., & Khandelwal, K. (2019). Artificial intelligence-based training learning from application. *Development and Learning in Organizations: An International Journal, 33*(2), 20–23. doi:10.1108/DLO-05-2018-0058

van der Goot, M. J., Hafkamp, L., & Dankfort, Z. (2021). *Customer Service Chatbots: A Qualitative Interview Study into the Communication Journey of Customers. Paper presented at the Chatbot Research and Design*, Cham.

Verganti, R., Vendraminelli, L., & Iansiti, M. (2020). Innovation and design in the age of artificial intelligence. *Journal of Product Innovation Management, 37*(3), 212–227. doi:10.1111/jpim.12523

Wang, D., & Li, V. J. (2019). Mass appraisal models of real estate in the 21st century: A systematic literature review. *Sustainability (Switzerland)*, *11*(24). doi:10.3390/su11247006

Westera, W., Prada, R., Mascarenhas, S., Santos, P. A., Dias, J., Guimarães, M., ... Ruseti, S. (2020). Artificial intelligence moving serious gaming: Presenting reusable game AI components. *Education and Information Technologies*, *25*(1), 351–380. doi:10.1007/s10639-019-09968-2

Woschank, M., Rauch, E., & Zsifkovits, H. (2020). A review of further directions for artificial intelligence, machine learning, and deep learning in smart logistics. *Sustainability (Switzerland)*, *12*(9). doi:10.3390/su12093760

Wunder, S. (2013). When payments for environmental services will work for conservation. *Conservation Letters*, *6*(4), 230–237. doi:10.1111/conl.12034

Yang, G., Ji, G., & Tan, K. H. (2022). Impact of artificial intelligence adoption on online returns policies. *Annals of Operations Research*, *308*(1), 703–726. doi:10.1007/s10479-020-03602-y

Zhu, X., Zhang, G., & Sun, B. (2019). A comprehensive literature review of the demand forecasting methods of emergency resources from the perspective of artificial intelligence. *Natural Hazards*, *97*(1), 65–82. doi:10.1007/s11069-019-03626-z

Chapter 3

ML techniques for analyzing security threats and enhancing sustainability in medical field based on Industry 4.0

Sukhpreet Kaur Khalsa
University of Alberta, Edmonton, Canada

Ranjodh Kaur
Guru Nanak Dev Engineering College, Ludhiana, India

Rajwinder Kaur
Sri Guru Granth Sahib World University, Fatehgarh Sahib, India

3.1 INTRODUCTION

Healthcare IoT encountered major changes in the healthcare industry. Before ML, patients were limited to visiting, calling, or texting to communicate with the doctor. As a result, specialists or emergency clinics could not screen patients' health consistently and make recommendations properly. Nowadays, remotely monitoring a patient's well-being is possible, including other characteristics like remote treatment using robotic surgeries and tele-auscultation (Valdez et al., 2021). As per the studies related to Industry 4.0, medical manufacturing organizations have enhanced their efficiency by 82% due to the assistance of Industry 4.0 technologies AI and IoT.

As the medical industry includes much sensible information, there is a need to design more secure and sustainable frameworks. To conduct a remote surgery, a high data rate is needed. Direct attacks against linked surgical robots and indirect attacks against connected equipment are two categories into which security threats can be classified (Koutras et al., 2020). If an operator resides within the remote site's proximity, he can reduce or mitigate the function degradation resulting from large latencies (Ullah et al., 2019). 5G networks would be able to handle a higher volume of data while maintaining reliability and reducing latency problems, improving access to mobile robotic surgery (Latif et al., 2017). To handle the security concerns of IoMT and enhance the sustainability in the IoT feature of Industry 4.0, there is a need to implement advanced techniques of ML.

DOI: 10.1201/9781003453567-3

Artificial intelligence, known as machine learning, learns from information and experiences without being explicitly programmed. For repeated data analysis and creating actionable information, ML can be a key component of the IoMT, particularly at processing nodes like cloud and fog computing (Jena et al., 2019). Recently, fascinating ML applications in IoT and IoMT have developed. To address IoMT security issues, the primary categories of ML algorithms can be used, named supervised, unsupervised, and semi-supervised (Hossen et al., 2022).

To increase sustainability in manufacturing, the concepts of deep learning (DL) and green technology can be used, which would be helpful to increase the security of machines at the core as well (Kováčová and Lăzăroiu, 2021). For processing and analyzing manufacturing data, DL-based models and the development of DL techniques are viewed favorably as a means of delivering sophisticated analytics capabilities. DL-based methods can also improve an industry's sustainability performance (*Jamwal et al., 2022*). Green computing is a key idea that can potentially transform every aspect of digital life. There are numerous strategies, plans, and methods for reducing energy consumption, but recycling and reusability are two important problems that can be resolved with end users' knowledge and developers' intelligence. Major green computing strategies, a comparison of various green computing technologies, and industry implementations have been elaborated by Kaur and Dhindsa (2015).

3.1.1 Research methodology

The process of study follows the various features of the approach described by Denyer and Tranfield (2009) and answers the following predefined research questions:

- RQ1: Which technologies of Industry 4.0 are used in the IoMT?
- RQ2: What are the security threats available?
- RQ3: What features of ML techniques can become a solution to the present problems?
- RQ4: What other solutions are available and can be integrated with ML techniques to secure the IoMT industry and increase sustainability?

3.2 MEDICAL INDUSTRY AND TECHNOLOGIES OF INDUSTRY 4.0

The healthcare industry is a popular field where multiple applications have developed for the past two years. Most applications are used for keeping fitness-related records, and a growing number of records are available. As the features of Industry 4.0, AI, and IoT have a significant role in the designing and development phase of the medical machines manufacturing field, AI and IoT healthcare devices are mainly focused on the following factors:

- Make healthcare faster, more personal-centered, and more affordable
- Focused on the fitness of the affected person
- Assist health experts during their duty

IoMT is the field where the healthcare system can be integrated with IoT. An integrated framework designed by Kaur and Dhindsa (2022) shows how the vast area of the medical field can be used for designing embedded systems using smartphones and the technologies of Industry 4.0 to enhance security and sustainability. Nowadays, healthcare using IoT is not limited to wearable devices, which are used to monitor the fitness of health, but it has been enhanced to find the solution to the physical presence of doctors while surgery as well. Recent applications of IoT healthcare are listed as follows:

- Tracking real-time location
- Monitoring hand hygiene
- Remote health monitoring
- Ambient and specially abled people in assisted living
- Assistance to specially abled people
- Remote/robotic surgery
- Smart labeling on drugs
- Screen hand sterility (RFID-labeled blood containers)

3.3 EVOLVED SECURITY THREATS FOR IOMT IN INDUSTRY 4.0

A means of defending something against potential injury or any other undesired forceful change is known as security. Others may harm you through a variety of techniques. Security threats can be represented by things and institutions, ecosystems, people and social groups, or anything susceptible to unwelcome change.

Cybersecurity and IT security are other well-known terms for computer security. Computer networks, smartphones, laptops, other computing devices, and the internet are all included. It covers safeguarding the tools used to access systems and the safeguarding of the hardware, software, data, and people. Physical security and information assurance are two issues that cybersecurity experts address with the aid of physical and virtual security (Hameed et al., 2021).

IoT security covers the security of electronic devices and computers on a single platform. It alludes to many security measures used for network- or internet-connected devices. Three fundamental factors, referred to as CIA, govern network security (confidentiality, integrity, and availability):

- Confidentiality: Only access by people who are authorized
- Integrity: Information is true to what it should be
- Availability: Information can be accessible when it needs to

Furthermore, security is based on triple-A, where authentication checks identity validation. It has three factors known what users know (user name/password), what users have (phone code, any token), and who are users (biometric/thumbprint/face recognition). Authorization describes what users can do after authentication, and accounting defines what happened and what a user did after authentication and authorization.

The fusion of the medical, electronic, and computer industries is called the "Internet of Medical Things" technology. Because IoMT directly affects individuals, hospital workers, and patients' lives, security is a key component. The data and information presented here are quite delicate. More danger can result from a single weakness because the alteration, publication, or absence of essential information may be fatal. Therefore, ensuring the connection between IoMT systems at all tiers of its design must be important.

A segment of the IoT network consists of IoMT devices. IoMT devices rely on IoT-specific protocols, standards, and technologies for communication. However, not all of them are utilized by IoMT devices. The IoT's three-layered architecture is the foundation for the list of protocols, standards, and technologies. Table 3.1 describes the protocols/technologies/standards used in the context of the IoMT industry.

When designing and developing the most popular protocols used in IoMT, various important security precautions must be considered, but before doing so, there is a need to understand the types of security threats and their impacts. Table 3.2 describes security vulnerabilities and threats in all protocols of the IoMT industry.

Table 3.1 Description of IoMT architecture's layers.

S. No.	Layer	Protocols/Technologies/Standards
1	Perception Layer	IEEE 802.15.4 standard, Infrared, RFID, NFC, Bluetooth/ BLE, Z-Wave, UWB
2	Network Layer	IEEE 802.15 standard, Wi-Fi, ZigBee, WIA-PA, 6LoWPAN, LoRaWAN
3	Application Layer	HL7, COAP, MQTT, HTTP

Table 3.2 Security issues and possible attacks in the IoMT industry (Koutras et al., 2020).

S. No.	Layer	Protocols/ Technologies/ Standards	Security Issues (SI) and Possible Attacks (PA)
1	Perception Layer	Infrared	• SI: No embedded security controls available in IR, works in very close proximity only • PA: Attackers can intercept the IR beam, snoop on data transmitted
		RFID	• SI: Embedded data are read-only and unsecured, with no by default authentication controls against tag scanning • PA: Attacks on the privacy of equipment, devices, and medical data; unauthorized cloning or tracking of tags; replay and DoS attacks
		NFC	• SI: Lack of strict security measures to prevent close-range attacks • PA: Man-in-the-Middle, Denial-of-Service, Bit manipulation
		BLE	• SI: Encrypts the payload only. Matching frequency hops is possible • PA: Firmware modification, software injection using Hciconfig software, sniffing and capturing of packets
		Z-Wave	• SI: Lack of standard key exchange protocol, trust in MPDU aggregation frame • PA: Key reset, impersonation attacks, node spoofing, black hole
		UWB	• SI: Long symbol length present, vulnerability to the access control list • PA: Early detection and late commit, same-nonce
2	Network Layer	Wi-Fi	• SI: In all networks, WPA2 is not implemented by default, Insufficiently detailed device authentication, MAC spoofing • PA: Data can be easily decrypted, DoS, spoof the MAC address, node tampering, proximity attacks, channel collision
		ZigBee	• SI: Implementation and protocol vulnerabilities, insufficient network key registration, insufficient verification of PAN IDs • PA: Easily hacking, key compromise, reset to factory
		WIA-PA	• SI: Absence of public-key encryption, absence of intrusion detection system, and broadcast key • PA: Sybil, wormhole, jamming, DoS, and traffic analysis

(Continued)

Table 3.2   (Continued)

S. No.	Layer	Protocols/ Technologies/ Standards	Security Issues (SI) and Possible Attacks (PA)
		6LoWPAN	• SI: IP N/W and radio signal vulnerabilities • PA: DoS, signal jamming
		LoRaWAN	• SI: Absence of re-keying, fake gateway signals, using past messages • PA: Password hacking, replay attack, battery exhaustion, message modifications
3	Application Layer	HL7	• SI: Message sources validation, unverified HL7 message sizes • PA: Spoofing, flooding
		COAP	• SI: DTLS implementation • PA: Parsing, cache, amplification, and cross-protocol attacks
		MQTT	• SI: Absence of encryption, IP broker • PA: Hacking, port obfuscation, packet analysis, and Botnet over MQTT
		HTTP	• SI: Absence of encryption, GET Method • PA: Eavesdropping, waste, GET, and HTTP flood

3.3.1 Possible solutions to security threats for IoMT in Industry 4.0

Various measures can be followed to enhance the security of the technologies of Industry 4.0 in IoMT. Advanced cipher combinations and ML algorithms are defined in this section as a solution to the possible security threats.

Firstly, to secure data in transit, it is necessary to know the answers to the following questions:

• Why do we secure data in transit?
• What are we securing?
• Where do we secure it?
• When do we secure it?
• Who secures it?
• How do we secure it?

There are different methods present to secure data in transit:

• Virtual private networks
• Remote access
• Site-to-site
• IPSec

- Transport layer security (TLS)
- The onion router (ToR)

The TLS cipher suite has different components to secure data in transit. The most secure combination of TLS cipher suite components to date is:
"TLS_ECDHE_RSA_WITH_AES_128_GCM_SHA256"
This combination includes the following components:

- Cipher suite: TLS1.2
- Key exchange: ECDHE
- Digital signature (authentication): RSA
- Data encryption: AES_128_GCM
- Message integrity: SHA256

The relation between the medical field, technologies of Industry 4.0, and ML techniques can be described systematically to define the concept of security. Hence, Figure 3.1 depicts the relationship between the IoMT industry and ML techniques for security.

As shown in Figure 3.1, supervised ML can secure implantable and wearable devices like smartwatches and smart fitness. The supervised ML approach works with data with known labels or classes. Regression and classification are two types of this approach. Neural networks, support vector machine, and decision Trees are the techniques that can prevent IoMT devices from various anomalies on different layers. The main purpose of these techniques is to detect attacks and malware.

Unsupervised ML can help classify the label based on its shared characteristics. Self-organizing map (SOM), K-nearest neighbor (KNN), and Latent Dirichlet Allocation are the major techniques that can be used to enhance the efficiency of detection of vulnerabilities. Moreover, these techniques work well for identifying new threats that exhibit anomalies.

In semi-supervised learning, a more recent branch of machine learning, certain learning data are labeled. These models can also detect threats and prevent hostile attacks on ML algorithms to prevent IoT devices from attacks.

A new area of machine learning called DL, a sophisticated type of neural network, has recently evolved in addition to the core classes of ML. It has numerous layers of artificial neural networks that replicate how the human brain processes information for object detection, obstacle detection, speech recognition, language translation, and decision-making. Hence, ML techniques can be added to enhance the security and sustainability of the IoMT industry. Figure 3.2 depicts the different areas of IoMT where ML methods can be employed.

As shown in Figure 3.2, ML techniques can be applied in the five major areas to detect and control the security vulnerabilities of the IoMT industry. The most popular mechanism for sensor security is machine learning for anomaly recognition.

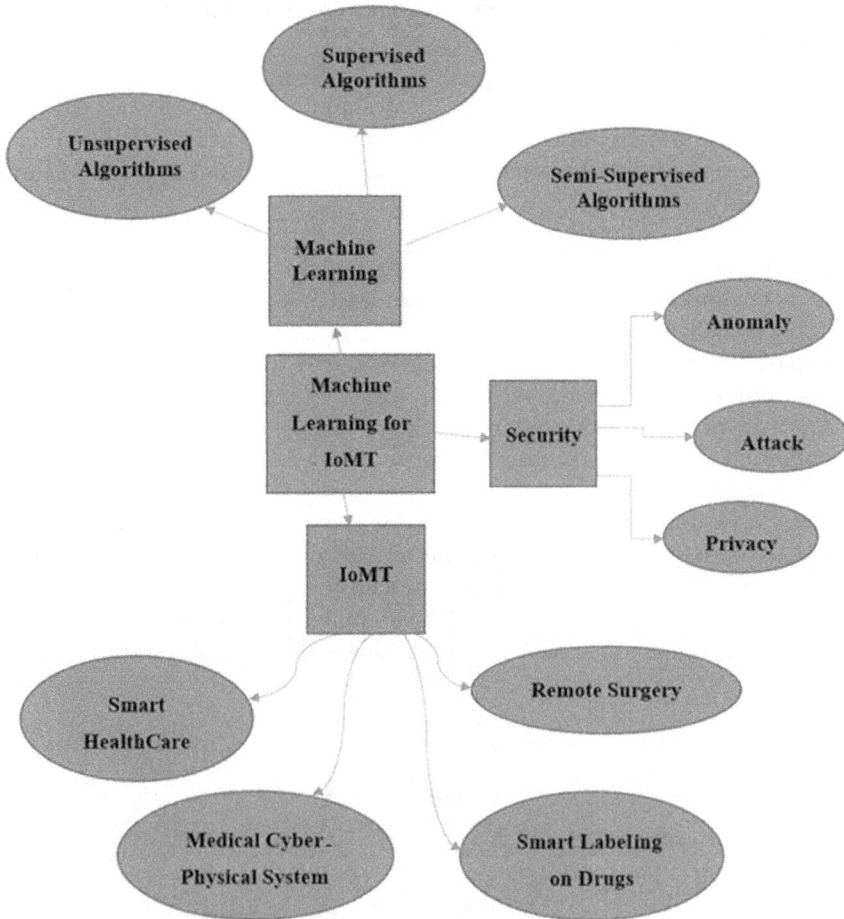

Figure 3.1 Relation of medical field, Industry 4.0, and ML.

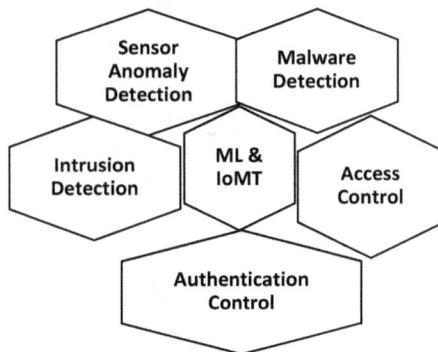

Figure 3.2 ML and area of IoMT industry.

3.4 APPROACHES TO SUSTAINABILITY FOR IOMT IN INDUSTRY 4.0

Sustainability has a significant role in designing and developing environment-friendly systems during manufacturing in Industry 4.0. Figure 3.3 depicts the relationship between the techniques of Industry 4.0 in IoMT and sustainability.

In Figure 3.3, suitability in Industry 4.0 depends on various factors. Adaptive manufacturing can reduce waste, increase time efficiency, and help in mass customization. Virtualization can lead to cost reduction and remote maintenance. System integration is a way to reduce the surplus and increase transparency. Simulation of manufacturing can improve the process and lead to predictive maintenance.

Apart from that, the concept of green technology might help in enhancing sustainability during manufacturing in Industry 4.0. Both green computing practices and IT users' perceptions of their intended beliefs have been studied by Chow and Chen (2009). The theories of planned behavior and reasoned action have supported the subjective norm for attitude. The perceived consequence of behavioral control concerning green computing directly impacts intention. Environmental sustainability can be achieved by "green computing," in which no component of the machines pollutes the

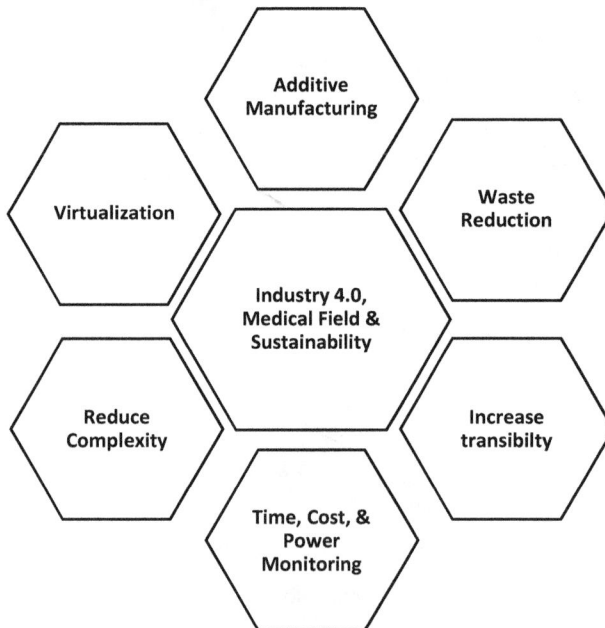

Figure 3.3 Sustainability and Industry 4.0.

environment and reduces harmful material used in machines, boosts energy efficiency, and creates recyclability of waste. Green computing includes the following four concepts:

- Green manufacture
- Green consumption
- Green disposal
- Green design

Table 3.3 includes the various studies showing how sustainability can be achieved by using ML during manufacturing for Industry 4.0 techniques.

As a result, IoMT may apply a variety of ML, green technology, and DL methodologies to improve the sustainability of Industry 4.0 manufacturing.

Table 3.3 Various studies on sustainability in Industry 4.0.

S. No.	Title	Outcome	Authors
1	"Deep learning for Manufacturing Sustainability: Models, Applications in Industry 4.0 and Implications"	A proposed framework to enhance sustainability	Jamwal et al., 2022
2	"Understanding the Adoption of Industry 4.0 Technologies in Improving Environmental Sustainability"	20 applications are identified and discussed to increase sustainability in Industry 4.0	Javaid et al., 2022
3	"The Sustainable Manufacturing Concept, Evolution and Opportunities within Industry 4.0: A Literature Review"	Discussed theoretical and empiricalframeworks	Sartal et al., 2020
4	"Blockchain-empowered Sustainable Manufacturing and Product Lifecycle Management in Industry 4.0: A Survey	Suitability through blockchain	Leng et al., 2020
5	"Machine Learning in Predictive Maintenance towards Sustainable Smart Manufacturing in Industry 4.0"	Comparison of different ML techniques of PdM	Cinar et al., 2020

3.5 CONCLUSIONS

ML techniques in Industry 4.0 significantly impact the designing and development of flexible, lightweight, integrated, and environment-friendly systems. Manufacturing in the medical field using the techniques of Industry 4.0 has been increasing since the past decade. The chapter elaborates on the importance of security and privacy concerns in the IoMT industry. It includes security issues based on each layer of IoT architecture and considers the possible attack correspondence to all included security issues. The relationship between IoMT and ML techniques in security solutions has been elaborated further. ML techniques like supervised, unsupervised, semisupervised, and deep learning are defined as security measures. Approaches to boost manufacturing sustainability in Industry 4.0 using ML are also elaborated.

The study concludes that various security concerns exist at different levels of manufacturing of IoMT devices. Combining TLS cipher and ML algorithms can cope with security issues. Moreover, sustainable manufacturing of security prevention techniques and tools through ML, DL, and green technology (GT) can reduce privacy concerns and provide environment-friendly manufacturing to the IoMT in Industry 4.0.

3.5.1 Future scope

Various devices have been designed for IoMT to improve the way of patient treatment. As the chapter discusses the present security threats in medical machines for different reasons, dealing with all these concerns based on the proposed solution can be considered a scope for the future. Furthermore, designing an integrated framework based on the discussed technologies, such as ML algorithms, DL, GT, and cipher suits for manufacturing secure and sustainable machines for Industry 4.0, is also possible.

REFERENCES

Chow, W.S., & Chen, Y. (2009). Intended belief and actual behavior in green computing in Hong Kong, *Journal of Computer Information Systems*, 50(2), 136–141. doi: 10.1080/08874417.2009.11645392

Cinar, Z.M., Nuhu, A.A., Zeeshan, Q., Korhan, O., Asmael, M., Safaei, B. (2020). Machine learning in predictive maintenance towards sustainable smart manufacturing in Industry 4.0, *Sustainability*, 12(19), 1–42. doi: 10.3390/su12198211

Denyer, D., & Tranfield, D. (2009). Producing a systematic review. In *The SAGE Handbook of Organizational Research Methods*, Thousand Oaks, CA, USA, Sage Publications Ltd, 671–689. https://www.cebma.org/wp-content/uploads/Denyer-Tranfield-Producing-a-Systematic-Review.pdf

Hameed, S.S., Hassan, W.H., Latiff, L.A., & Ghabban, F. (2021). A systematic review of security and privacy issues in the internet of medical things; the role of machine learning approaches, *Peer J Computer Science*, 7(4), 1–44. doi: 10.7717/peerj-cs.414

Hossen, M.N., Panneerselvam, V., Koundal, D., Ahmed, K., Bui, F.M. & Ibrahim, S.M. (2022). Federated machine learning for detection of skin diseases and enhancement of Internet of Medical Things (IoMT) security, *IEEE Journal of Biomedical and Health Informatics*, 1–1. doi: 10.1109/JBHI.2022.3149288

Jamwal, A., Agarwal, R., & Sharma, M. (2022). Deep learning for manufacturing sustainability: Models, applications in Industry 4.0 and implications, *International Journal of Information Management Data Insights*, 2(2), 1–13. doi: 10.1016/j.jjimei.2022.100107

Javaid, M., Haleem, A., Singh, R.P., Suman, R., & Gonzalez, E.S. (2022). Understanding the adoption of Industry 4.0 technologies in improving environmental sustainability, *Sustainable Operations and Computers*, 3, 203–217. doi: 10.1016/j.susoc.2022.01.008

Jena, M.C., Mishra, S.K., & Moharana, H.S. (2019). Application of Industry 4.0 to enhance sustainable manufacturing, *Environmental Progress & Sustainable Energy*, 39, 1–11. doi: 10.1002/ep.13360

Kaur, S., & Dhindsa, K.S. (2015). Green computing - saving the environment with intelligent use of computing. *CSI Communication*, 39(5), 16–18. www.csi-india.org

Kaur, S., & Dhindsa, K.S. (2022). IFSA: an integrated framework for developing IoT linked mobile applications for specially abled people, *Wireless Networks*, 28(4). doi: 10.1007/s11276-022-02905-1

Koutras, D., Stergiopoulos, G., Dasaklis, T., Kotzanikolaou, P., Glynos, D., & Douligeris, C. (2020). Security in IoMT communications: A survey, *Sensors*, 20(4828), 1–49. doi: 10.3390/s20174828

Kováčová, M., & Lăzăroiu, G. (2021). Sustainable organizational performance, cyber-physical production networks, and deep learning-assisted smart process planning in Industry 4.0-based manufacturing systems. *Economics, Management, and Financial Markets*, 16(3), 41–54. https://www.ceeol.com/search/article-detail?id=983513

Latif, S., Qadir, J., Farooq, S., & Imran, M.A. (2017). How 5G (and concomitant technologies) will revolutionize healthcare. *Future Internet*, 9(93), 1–24. doi: 10.48550/arXiv.1708.08746

Leng, J., Ruan, G., Jiang, P., Xu, K., Liu, Q., Zhou, X., and Liu, C. (2020). Blockchain-empowered sustainable manufacturing and product lifecycle management in industry 4.0: A survey. *Renewable and Sustainable Energy Reviews*, 132, 1–20. doi: 10.1016/j.rser.2020.110112

Sartal, A., Bellas, R., Mejias, A.M., & Garcia-Collado, A. (2020). The sustainable manufacturing concept, evolution and opportunities within Industry 4.0: A literature review. *Advances in Mechanical Engineering*, 12, (5), 1–17. doi: 10.1177/1687814020925

Ullah, H., Nair, N.G., Moore, A., Nugent, C., Muschamp, P., & Cuevas, M. (2019). 5G communication: An overview of vehicle-to- everything, drones, and healthcare use-cases. *IEEE Access*, 7, 37251–37268. doi: 10.1109/ACCESS.2019.2905347

Valdez, L.B., Datta, R.R., Babic, B., Müller, T.D., Bruns, J.C., & Fuchs, F.H. (2021). 5G mobile communication applications for surgery: An overview of the latest literature. *Artif Intell Gastrointest Endosc*, 2(1), 1–11. doi: 10.37126/aige.v2.i1.1

Chapter 4

Role of machine learning in cyber-physical systems to improve manufacturing processes

Raman Kumar, Sita Rani, and Sehijpal Singh

Guru Nanak Dev Engineering College, Ludhiana, India

4.1 INTRODUCTION

Manufacturing procedures are now more sophisticated and demanding than ever before. How things are created, developed, and manufactured has substantially altered with the introduction of cyber-physical systems (CPS) in manufacturing processes. CPS create a linked network of devices that can interact with each other to optimize manufacturing processes by fusing physical processes with digital technology (Babiceanu & Seker, 2016).

In CPS, machine learning (ML) has become a potent tool. The insights ML algorithms can extract from data may improve manufacturing. Large volumes of data may be analyzed using ML to assist manufacturers in finding patterns, forecast outcomes, and optimizing operations to increase productivity, save costs, and improve product quality (Ahmed, Ahmed, & Saeed, 2021).

This chapter provides an overview of the role of ML in CPS to improve manufacturing processes. It discusses on various ML techniques, their applications, and the challenges of implementing ML in manufacturing. The chapter also features case studies highlighting successful implementations of ML across multiple manufacturing industries. Finally, it explores future directions and opportunities for ML in manufacturing processes and concludes with a summary of key findings, implications for the future of manufacturing processes, and recommendations for practitioners and decision-makers.

4.2 EXPLANATION OF CYBER-PHYSICAL SYSTEMS IN MANUFACTURING PROCESSES

In manufacturing processes, the term "CPS" refers to integrating digital technology like sensors, software, and communication networks with physical systems like machinery and equipment. Using digital technology for real-time monitoring and control of physical systems, CPS enables manufacturers to optimize their processes for optimal effectiveness, productivity,

DOI: 10.1201/9781003453567-4

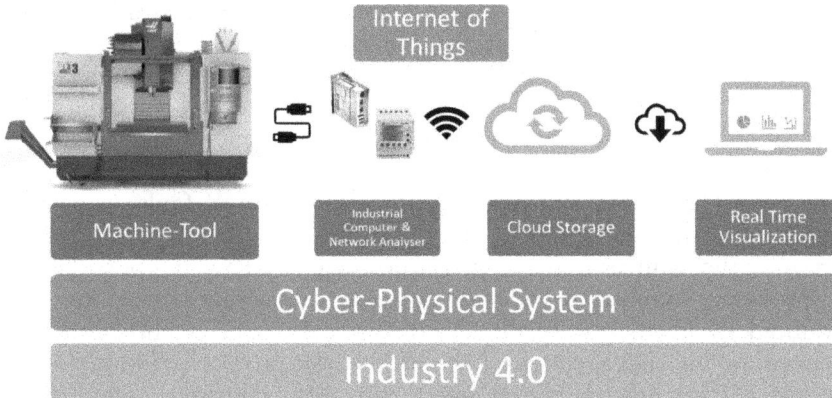

Figure 4.1 Overview of cyber-physical system, adapted from Morella et al. (2020) by CC 4.0.

and quality. Manufacturing systems may become more sensitive to changes in demand, supply chain interruptions, and other outside variables. CPS can give insights into the manufacturing process that can be utilized to optimize operations and save costs by integrating data analytics and ML algorithms (Ma, Zhang, Lv, Yang, & Wu, 2019). An overview of CPS is shown in Figure 4.1. It shows the connectivity of machine tools, industrial computer, and network with cloud storage and real-time visualization (Morella, Lambán, Royo, Sánchez, & Ng Corrales, 2020).

Using sensors to monitor equipment performance, ML algorithms to streamline production, and robotics and automation to boost output and save labor costs are a few examples of manufacturing CPS. CPS can also provide real-time supply chain monitoring and management, enabling manufacturers to respond quickly to changes in demand or issues with the supply chain. CPS may alter the manufacturing sector by boosting productivity, cutting costs, and improving product quality. By making better use of the massive amounts of data generated by manufacturing processes, integrating CPS and ML algorithms will assist manufacturers in developing more intelligent and productive production systems (Zhang, Wang, Ding, Chan, & Ji, 2020).

4.3 SIGNIFICANCE OF MACHINE LEARNING IN MANUFACTURING PROCESSES

Due to its capacity to evaluate and learn from enormous volumes of data generated by manufacturing systems, ML has grown in significance in manufacturing processes. Manufacturers may acquire insights into the production process by utilizing ML algorithms that would be challenging or impossible to achieve using conventional approaches. The capacity of ML to increase efficiency and cut costs in manufacturing processes is one of its main advantages. By finding patterns, forecasting outcomes, and suggesting process modifications, ML systems may enhance manufacturing processes. For instance, manufacturers may use ML to spot trends in equipment failure rates, allowing them to carry out proactive maintenance and cut downtime (Angelopoulos et al., 2020).

By identifying flaws and lowering the production of faulty goods, ML may help raise product quality. ML algorithms may analyze data from sensors and other sources to find patterns pointing to a flaw or a possible quality problem. Manufacturers may decrease the number of faulty items produced by identifying quality problems early in production. Real-time monitoring and management of the production process is another advantage of ML in manufacturing operations. ML algorithms may identify changes in the production process and make modifications to maximize efficiency and quality by evaluating data in real time (Ranjan, Kumar, Kumar, Kaur, & Singh, 2022). Customer satisfaction may eventually increase due to quicker response times to changes in demand and supply chain interruptions. By providing real-time monitoring and control, increasing productivity and lowering costs, and improving product quality, ML offers a vast potential to enhance manufacturing processes. The significance of ML in manufacturing processes will only grow as the volume of data produced by manufacturing processes rises (Dogan & Birant, 2021).

4.4 CHALLENGES IN IMPLEMENTING MACHINE LEARNING IN MANUFACTURING PROCESSES

Although applying ML to manufacturing processes has many potential advantages, many obstacles must be overcome before it can be successfully used. The following are some of the significant difficulties in integrating ML into manufacturing processes (Fahle, Prinz, & Kuhlenkötter, 2020):

- Data quality: The efficacy of ML algorithms depends on the caliber of the data used to train them. Data in manufacturing may be inconsistent, erroneous, or incomplete, which can harm how well ML algorithms function.
- Data availability: In some circumstances, there may not be enough data to train ML algorithms fully. This might make it difficult to create precise models to examine and learn from the data.
- Integration with current systems: Integrating new ML techniques into manufacturing processes is frequently necessary. This could not be easy if methods and procedures weren't created with ML in mind (Kang, Catal, & Tekinerdogan, 2020).
- Resources and expertise: ML implementation in manufacturing processes requires manufacturing and ML knowledge. It's possible that

many firms don't have the knowledge or resources needed to integrate ML into their production processes properly.

- Explainability and transparency: It might not always be clear how ML algorithms decide what to do. This lack of openness may make it difficult to justify actions to stakeholders, such as regulators or consumers.
- Security and privacy: The application of ML to manufacturing procedures may entail delicate data, such as client information or trade secrets. It is crucial to guarantee the privacy and security of this data (Rani et al., 2022).

Consideration of these and other issues is necessary when implementing ML in manufacturing processes. Yet, with the appropriate strategy, ML can offer essential advantages in increased productivity, decreased costs, and improved product quality (Carvalho et al., 2019).

4.5 CASE STUDIES

4.5.1 Predictive maintenance in automotive manufacturing

A crucial use of ML in the production of automobiles is predictive maintenance. Manufacturers can reduce unexpected downtime and repair costs through proactive maintenance by foreseeing equipment faults before they happen. A significant carmaker is one example of how predictive maintenance has been successfully used in the automobile industry. The carmaker used ML algorithms to analyze sensor data from its production equipment, including robots and conveyors. The ML algorithms found patterns in the sensor data that showed early warning indicators of impending equipment breakdowns. Next, to provide maintenance workers enough time to make preventative repairs, the algorithms were used to forecast when equipment breakdowns were likely to happen. The carmaker used predictive maintenance, which resulted in a 10% reduction in equipment downtime and considerable cost savings (Theissler, Pérez-Velázquez, Kettelgerdes, & Elger, 2021). The carmaker might save expenses by proactively maintaining its equipment and extending its lifespan. Overall, this case study shows the effectiveness of ML in predictive maintenance in the production of automobiles. Manufacturers may identify early warning signals of equipment problems and conduct proactive maintenance by utilizing sensor data and ML algorithms, which lowers costs and boosts productivity (Cheng et al., 2022; Florian, Sgarbossa, & Zennaro, 2021).

4.5.2 Quality control in semiconductor manufacturing

ML algorithms may play a significant role in assuring good product quality. Quality control is an essential component of the semiconductor manufacturing process. Leading semiconductor manufacturing is one instance of how ML has been successfully used in quality control. To find flaws and other quality problems in photos of semiconductor wafers, the firm deployed ML algorithms. The computers could find patterns that indicated certain flaws, such as scratches or particle contamination, by examining thousands of photos (D. Kim, Kang, Cho, Lee, & Doh, 2012). The ML algorithms then employed the existence of these patterns to categorize each wafer as either excellent or faulty. This made it possible for the producer to immediately spot flawed wafers and adopt remedial measures to lower the production of faulty goods. The semiconductor firm significantly saved costs by reducing the number of damaged wafers by 10% by adopting ML in quality control. Moreover, the manufacturer could limit the flaws that made it into the finished product by recognizing quality concerns early in the production process, further enhancing product quality. Ultimately, this case study shows how ML may improve quality control in the semiconductor industry. Manufacturers can immediately spot quality problems and take remedial action, improving product quality and lowering costs by analyzing photos and spotting trends (S. H. Kim, Kim, Seol, Choi, & Hong, 2022).

4.5.3 Process optimization in food and beverage manufacturing

Manufacturing food and beverages requires careful process optimization, and ML techniques may significantly increase productivity and save costs. A significant food and beverage business is one example of how ML has been successfully used in process improvement. The firm analyzed manufacturing process data using ML techniques, including details on temperature, pressure, and other elements. The computers found trends in this data that pointed to the potential for process optimization, such as modifying cooking durations or ingredient ratios (M. Li & Li, 2022). The production processes were subsequently optimized using ML algorithms, leading to considerable increases in productivity and cost reductions. For instance, the producer increased output capacity by 10% by cutting cooking times by 5%. In addition, the firm decreased material waste by 8% by improving ingredient ratios. The maker of food and beverages boosted productivity, saved expenses, and enhanced efficiency by integrating ML into process optimization. These upgrades raised profitability and assisted the firm in satisfying the rising demand for their goods. Overall, this case study illustrates the effectiveness of ML for process optimization in the production of food and beverages. Manufacturers may boost productivity, decrease costs, and increase efficiency by evaluating data and finding chances for improvement, resulting in considerable business benefits (Khan, Sablani, Nayak, & Gu, 2022).

4.5.4 Supply chain management in pharmaceutical manufacturing

Pharmaceutical production requires effective supply chain management, and ML algorithms may significantly increase the effectiveness of the supply chain while cutting costs. One prominent pharmaceutical company is one example of how ML has been successfully used to supply chain management. The firm analyzed data from its supply chain using ML algorithms, including details on supplier performance, inventory levels, and demand projections. The algorithms could find chances for supply chain optimization by analyzing this data, such as by optimizing inventory levels or finding substitute suppliers. The supply chain was optimized using ML algorithms, significantly increasing efficiency and reducing costs. As an illustration, the manufacturer cut inventory levels by 15%, which resulted in a 5% decrease in inventory expenses. Also, the firm was able to save 10% on procurement expenses by locating alternative suppliers. The pharmaceutical producer boosted agility, saved costs, and improved supply chain efficiency by integrating ML into supply chain management. These upgrades raised profitability and made it easier for the firm to satisfy customer demand for their goods. Ultimately, this case study shows how ML may be used to manage the supply chain for pharmaceutical manufacturing. Manufacturers may boost the effectiveness of their supply chains, save costs, and increase agility, all of which positively impact their bottom line (Naz et al., 2022; Tirkolaee, Sadeghi, Mooseloo, Vandchali, & Aeini, 2021).

4.6 BEST PRACTICES FOR IMPLEMENTING MACHINE LEARNING IN MANUFACTURING PROCESSES

4.6.1 Identify clear business objectives

Clearly defining company goals is crucial when integrating ML into manufacturing processes. These goals must be clear, quantifiable, doable, pertinent, and time-bound (SMART). By doing this, manufacturers may ensure that their ML implementation aligns with their corporate objectives (Shaikh, Shinde, Rondhe, & Chinchanikar, 2023). To identify clear business objectives, manufacturers should ask themselves the following questions:

- What specific problem or challenge do we want to solve with ML?
- How will ML help us achieve our business goals?
- What metrics will we use to measure the success of our ML implementation?
- What resources do we need to implement ML effectively?

Manufacturers may ensure that their ML implementation goals and objectives align with their business goals and objectives by providing unambiguous answers to these questions. Also, it's crucial to include significant stakeholders while defining corporate objectives. Stakeholders from many departments within the company, such as the manufacturing, engineering, and quality control teams, are included in this. Manufacturers may ensure their aims are pertinent and reachable by incorporating essential stakeholders. Defining explicit business objectives is the first step in successfully adopting ML in manufacturing processes. Manufacturers may guarantee that they align with their corporate aims and objectives and position themselves for success by doing this (Rai, Tiwari, Ivanov, & Dolgui, 2021).

4.6.2 Build a cross-functional team

Collaboration across several teams and departments, including manufacturing, engineering, quality control, and IT, is necessary for applying ML. A cross-functional team should comprise people with a range of experiences and qualifications. For instance, data scientists, engineers, production managers, and business analysts could be on the team. Due to its diversity, the team can better view problems from various perspectives and find solutions. The team should also have a distinct leader in applying the ML. The business goals and technical components of ML deployment should be well understood by this leader. They have to be able to interact with stakeholders productively and guarantee that everyone on the team is focused on the same objective. Involving stakeholders from many departments in the deployment of ML is also crucial (Bäuerle et al., 2022). This comprises the decision-makers and end users whom the ML implementation will impact. The cross-functional team may ensure that the ML implementation aligns with the organization's requirements and fulfills end users' expectations by including these stakeholders. A cross-functional team must be established for ML to use in manufacturing processes successfully. The team may ensure that the ML deployment aligns with business objectives and addresses end-user demands by assembling people with different backgrounds and skill sets and involving key stakeholders (Terranova, Venkatakrishnan, & Benincosa, 2021).

4.6.3 Establish data governance and management processes

Data accuracy, consistency, and security are ensured through efficient ML application through data governance and management protocols. The following procedures should be taken into consideration by manufacturers when establishing data governance and management processes (Janssen, Brous, Estevez, Barbosa, & Janowski, 2020):

- Specify the criteria for data quality: Data quality standards that specify what counts as accurate, complete, and consistent data should be established by manufacturers. The ML implementation's business goals should align with these criteria.

- Create data management procedures: Manufacturers must create procedures for managing data at every stage of its lifespan, from data collection to destruction. Data integration, storage, and cleansing should all be part of these procedures.
- Ensure data is secure: Manufacturers must ensure data is safe and shielded from unwanted access. Implementing access restrictions, encryption, and other security measures falls under this category.
- Preserve data privacy: Manufacturers ensure data is gathered and used following all applicable privacy laws, including the CCPA and GDPR.
- Create a data governance team: Manufacturers should assemble a group monitoring the administration and governance of their company's data. Representatives from many departments should be on this team and be in charge of enforcing data governance regulations.

Manufacturers may ensure that they are gathering, managing, and using data to align with their business objectives and complying with applicable requirements by implementing data governance and management systems (Blomster & Koivumäki, 2022; Shin, Lee, & Hwang, 2020).

4.6.4 Develop a robust infrastructure for data collection and storage

Large amounts of high-quality data must be collected, stored, and managed by a solid infrastructure for training ML algorithms, which demands the development of a robust data collection and storage infrastructure. Manufacturers should think about taking the following actions to build a robust infrastructure for data collecting and storage (Gohel, Upadhyay, Lagos, Cooper, & Sanzetenea, 2020):

- Determine the data sources: The manufacturers should identify the data sources that will be used for ML implementation. This might consist of sensor data, machine records, and output information.
- Establish data collection procedures: Manufacturers should put data collection procedures in place to gather data from sources that have been identified. This might entail interfacing with current systems, adding sensors, or setting up data-collecting software.
- Effectively store data: Manufacturers must ensure that data is adequately kept, with adequate backups and disaster recovery strategies. This can entail leveraging cloud-based storage options or establishing a data warehouse.
- Assure high data quality: Manufacturers should implement data cleansing and validation systems to guarantee high data quality.
- Infrastructure should be scaled as needed to accommodate increasing data volumes. Manufacturers should do this. This might entail setting up a distributed computing architecture, increasing storage capacity, or installing extra servers.

Manufacturers may ensure they are gathering and keeping high-quality data that can be utilized to train ML algorithms efficiently by building a solid data collection and storage infrastructure (Deist et al., 2017; Mitri et al., 2017).

4.6.5 Select appropriate machine learning techniques

Selecting the appropriate ML approach is crucial for producing accurate and reliable predictions since different ML techniques are suitable for various types of data and business challenges. The following measures should be taken into account by manufacturers when choosing suitable ML techniques (Khaledian & Miller, 2020; Luo, 2016):

- Establishing business goals: Manufacturers should define their corporate goals and the particular issues they are attempting to resolve. This will enable them to choose the best ML approach to solve those issues.
- Data analysis is necessary for manufacturers to choose the best ML approach. To choose the best ML approach, this investigation may entail finding patterns in the data or running statistical calculations.
- Choose the best ML approach: Based on the data analysis findings, manufacturers should choose the best technique. For instance, a decision tree or random forest approach may be suitable if the data contains categorical variables. A regression approach would be more appropriate if the data comprises continuous variables.
- Test and improve ML models: To make sure that ML models are reliable and efficient, producers should test and improve them. This might entail changing the methods, adding new data, or adjusting the ML algorithm settings.

Manufacturers may guarantee that they fulfill their business objectives and provide precise and effective predictions by choosing the proper ML approach (G. Li, Zhou, & Cao, 2021; Ray, 2019).

4.6.6 Continuously monitor and evaluate model performance

The quality of the training data might fluctuate over time. ML models are only as good as the data used to train them. To keep models accurate and efficient, continuous monitoring and assessment can assist in determining when they need to be updated or retrained. Manufacturers should think about taking the following actions to monitor and assess model performance regularly (Ehrlinger, Haunschmid, Palazzini, & Lettner, 2019):

- Specify the metrics for evaluation: Manufacturers need to specify evaluation measures to gauge how well their ML models work. These indicators must be straightforward to measure and tailored to the company's goals.

- Gather information for evaluation: Manufacturers must continuously gather information to assess their ML models' effectiveness. This information must be typical of the data used to train the models.
- Manufacturers should use performance indicators to assess the effectiveness of their ML models. This assessment should be done regularly, such as every day or every week.
- To make improvements, manufacturers must pinpoint the areas where their ML models fall short of expectations. This can need modifying the ML formulas or including new data.
- Retrain or update models: To keep their ML models accurate and efficient, manufacturers should retrain or update them as necessary. This might entail employing new algorithms or adding more data.

By regularly observing and assessing model performance, manufacturers can guarantee that their ML models continue to be precise and efficient over time (Kourouklidis, Kolovos, Matragkas, & Noppen, 2020; Rush, Celi, & Stone, 2019).

4.6.7 Foster a culture of innovation and continuous improvement

ML quickly changes, and new methods and tools are constantly being developed. Manufacturers may keep up with the most recent advancements in ML and continuously enhance their processes by cultivating an innovative culture. Manufacturers should think about taking the following actions to promote an innovation and continuous improvement culture (Costa et al., 2019):

- Promote trial and error: Manufacturers should encourage staff members to experiment with cutting-edge ML methods and tools. Proof-of-concept studies or pilot projects are two possible forms of this exploration.
- Training and education: To keep their staff members abreast of the most recent advances in ML, manufacturers should offer training and education to them. Workshops, seminars, or online courses may be a part of this training.
- Encourage creativity by rewarding staff members who create cutting-edge ML solutions. Examples include bonuses, promotions, and acknowledgments in corporate newsletters or other internal communication channels.
- Establish a culture that fosters innovation and ongoing development. Manufacturers should do this. This may entail assembling a unique team for innovation or organizing frequent workshops on innovation.

Manufacturers may stay on the cutting edge of the most recent breakthroughs in ML and continuously enhance their processes by cultivating a culture of innovation and continuous improvement (Qin & Chiang, 2019).

4.7 ETHICAL CONSIDERATIONS IN MACHINE LEARNING FOR MANUFACTURING PROCESSES

When applying ML to manufacturing processes, crucial ethical aspects must be considered, just like with any usage of artificial intelligence. These concerns cover matters like accountability, prejudice, and data privacy (Akinci D'Antonoli, 2020; Michelson, Klugman, Kho, & Gerke, 2022).

- Data protection: ML models need a lot of data to be adequate, some of which can contain private information about specific people. The privacy of the people whose data is being utilized must be protected, and it is crucial to ensure that data is gathered, kept, and used responsibly.
- Bias: ML models could be trained on biased data, providing biased results. The data used to train models should be carefully chosen, and bias should be checked regularly.
- Accountability is required when ML is used in manufacturing processes, just like with any other AI application. Manufacturers must be accountable for the outcomes produced by their models and open about how they use ML in their operations.

To address these ethical considerations, manufacturers should consider the following steps:

- Create explicit crystal standards about data privacy: Manufacturers should create regulations that specify how data will be gathered, kept, and utilized in ML models.
- Manufacturers should regularly check their ML models for bias and take action to correct any bias.
- Create accountability measures: Companies that use ML in their production processes should set up accountability processes, such as routine audits or reviews by impartial third parties.
- Interact with stakeholders: To ensure their use of ML is transparent and responsible, manufacturers should communicate with stakeholders, such as staff members, clients, and the general public.
- Create explicit crystal standards about data privacy: Manufacturers should create regulations that specify how data will be gathered, kept, and utilized in ML models.

Manufacturers may assist in guaranteeing that their use of AI is responsible, transparent, and accountable by considering these ethical factors when integrating ML into manufacturing processes. This may support the company's long-term success by helping to increase trust among stakeholders and clients (Rahmattalabi & Xiang, 2022; Russo, 2021).

4.8 FUTURE DIRECTIONS AND OPPORTUNITIES

4.8.1 Advancements in machine learning and cyber-physical systems

ML and CPS have a bright future in manufacturing, with many developments and possibilities to come. As ML and CPS develop, the manufacturing sector might dramatically transform due to increased productivity, less waste and downtime, and higher product quality. The followings are some of the most significant developments and changes in ML and CPS for manufacturing (Olowononi, Rawat, & Liu, 2021; Shishvan, Zois, & Soyata, 2018):

- Increased propensity for prediction: As ML algorithms advance, they will become ever more helpful in anticipating maintenance requirements, spotting quality problems, and streamlining manufacturing procedures.
- More automation: As CPS develops, it can automate more production processes, lowering the need for human involvement and boosting efficiency.
- Collaboration will increase as CPS becomes more networked, promoting better communication and cooperation between machines, people, and other systems and increasing production and efficiency.
- Improved product customization: ML and CPS will allow firms to adapt products to fit the tastes and demands of specific customers, boosting satisfaction and churn.
- Improved safety: ML and CPS will let producers monitor and regulate production processes in real time, lowering the likelihood of accidents and raising overall safety.
- Improved sustainability: ML and CPS will allow producers to optimize energy use, cut waste, and lessen the impact of manufacturing on the environment.

Manufacturers must keep up with the most recent advancements in ML and CPS and be ready to adopt new technologies and processes as these domains progress. By embracing these developments and possibilities, manufacturers may strengthen their processes, products, and position for long-term success in a quickly evolving sector (Nagarajan, Deverajan, Bashir, Mahapatra, & Al-Numay, 2022).

4.8.2 Potential impact on manufacturing processes

On manufacturing processes, ML and CPS have a lot of possible effects. Thanks to ML and CPS, manufacturing processes might undergo a revolution, boosting production, profitability, and efficiency. The following are some possible effects of ML and CPS on manufacturing processes (Radanliev, De Roure, Van Kleek, Santos, & Ani, 2021):

- ML and CPS may streamline manufacturing processes, reduce waste and downtime, and boost efficiency.
- Improved quality control: ML and CPS may be used to find and fix quality problems in real time, raising product quality.
- Enhanced automation: ML and CPS can automate various manufacturing processes, lowering the demand for human involvement and boosting productivity.
- ML and CPS may be used to forecast when maintenance is necessary, which lowers the possibility of unplanned downtime and raises equipment uptime.
- Increased customization: ML and CPS may be used to tailor products to fit the requirements and preferences of specific customers, boosting their happiness and loyalty.
- Improved safety: Manufacturing processes may be monitored and controlled in real time using ML and CPS, lowering the chance of accidents and raising overall safety.
- ML and CPS may be used to optimize energy utilization, cut waste, and lessen the environmental effect of manufacturing operations, leading to increased sustainability.

ML and CPS might significantly impact manufacturing processes, and if these technologies develop further, the influence is expected to grow even more. The manufacturers who adopt these technologies and change with the times will probably reap the most rewards for improved production, efficiency, and profitability (Calinescu, Cámara, & Paterson, 2019).

4.9 RESEARCH AND DEVELOPMENT OPPORTUNITIES

There are several chances for research and development in ML and CPS for manufacturing processes. These opportunities include the following, among others (Oliveira et al., 2021):

- Creating cutting-edge ML algorithms: More advanced ML algorithms are required so they can manage vast and complicated data sets and offer more precise predictions and insights.
- Integrating ML with other technologies: ML may be used in new applications and solutions for manufacturing processes, along with other cutting-edge technologies like the Internet of Things (IoT), blockchain, and augmented reality.

- Making ML applications for specific sectors: There is a need for industry-specific ML applications to meet these goals since different sectors have distinctive production processes and requirements.
- Creating ML applications for small- and medium-sized businesses (SMEs): Many SMEs lack the knowledge and resources necessary to implement ML solutions independently, so there is a need for more readily available and affordable ML solutions that can assist SMEs in streamlining their manufacturing processes.
- Privacy, prejudice, and job displacement are ethical and social challenges that must be addressed as ML and CPS become more common in manufacturing processes.
- Creating ML-enabled decision support systems: ML may be used to develop decision support systems that can assist engineers and managers in making better decisions in real time, increasing productivity, profitability, and efficiency.
- Creating ML applications for environmentally friendly manufacturing: ML may be used to improve energy efficiency, decrease waste, and lessen the environmental effect of manufacturing processes, resulting in more environmentally friendly manufacturing methods.

There are several potentials for innovation and growth in the research and development of ML and CPS for manufacturing processes, which is still in its early phases. Manufacturers can stay on the cutting edge of technological breakthroughs and help shape the industry's future by investing in research and development (Wickramasinghe, Marino, Amarasinghe, & Manic, 2018).

4.10 CONCLUSIONS

This chapter discusses the role of ML in improving manufacturing processes in the context of CPS. The chapter highlighted the significance of ML in manufacturing and how it can help manufacturers improve efficiency, reduce costs, and enhance product quality. The chapter identified challenges associated with implementing ML in manufacturing, such as data quality, privacy concerns, and the need for specialized skills and resources. This also presented best practices for implementing ML in manufacturing, such as identifying clear business objectives, building a cross-functional team, establishing data governance and management processes, and continuously monitoring and evaluating model performance. The chapter has discussed ethical considerations in using ML in manufacturing and the need to address issues such as bias, privacy, and job displacement. Finally, highlighted future directions and opportunities for research and development in ML and CPS for manufacturing processes, including developing advanced ML algorithms, combining ML with other technologies, and addressing social and ethical issues.

Overall, the use of ML in manufacturing processes shows excellent potential for improving efficiency, quality, and sustainability, and manufacturers who invest in this technology can gain a competitive advantage in the industry.

4.10.1 Implications for the future of manufacturing processes

The usage of ML in production processes will significantly impact the future of manufacturing. The use of ML to optimize manufacturing processes will be more and more crucial as firms continue to implement CPS and other cutting-edge technologies. The capacity of ML to increase productivity and efficiency in manufacturing processes is one of its main advantages. ML algorithms can find patterns and insights in enormous volumes of data from sensors and other sources that people might not be able to see. This may result in more effective manufacturing techniques, less waste, and higher-quality products.

By offering insights into demand forecasting, inventory management, and logistics, ML may assist firms in improving their supply chains. Ensuring that items are delivered on time and in the proper amounts may assist manufacturers in lowering costs and enhancing customer satisfaction. ML may significantly impact product design and development in addition to these operational advantages. Manufacturers may detect consumer requirements and preferences and create new goods that better satisfy those demands by evaluating customer data and feedback using ML algorithms. As a result, there may be an improvement in customer satisfaction, sales, and market competitiveness.

Future manufacturing processes are anticipated to be significantly impacted by ML as it develops further. For instance, combining ML with cutting-edge technologies like robots and the IoT may increase manufacturing productivity, automation, and efficiency. Applying ML to manufacturing processes can revolutionize the sector by enhancing productivity, sustainability, and quality while giving enterprises a competitive edge in the global market.

4.10.2 Final thoughts and recommendations

In conclusion, using ML in manufacturing processes offers enormous potential for enhancing the sector's effectiveness, productivity, and standard. ML algorithms can find patterns and insights by analyzing vast volumes of data from sensors and other sources, which enables businesses to improve their workflows and supply chains, create new products, and stay competitive in the global market. However, integrating ML into production procedures is not without its difficulties. In addition to fostering an innovative and continuous improvement culture, manufacturers must handle data governance, model selection, and performance monitoring difficulties.

Manufacturers should adhere to best practices like establishing clear business objectives, creating a cross-functional team, establishing data governance and management processes, creating a solid infrastructure for data collection and storage, choosing appropriate ML techniques, continuously monitoring and evaluating model performance, and encouraging a culture of innovation and continuous improvement to implement ML in manufacturing processes successfully. The ethical ramifications of ML in manufacturing, including data protection and security, bias and fairness, openness, and responsibility, must also be considered. By integrating ML into their operations, manufacturers should put ethical concerns first. As ML continues to develop, there will be new possibilities and difficulties for producers in the business. Manufacturers should keep up with ML developments and continually assess how they may use these technologies to enhance their operations and maintain competitiveness in the global market.

REFERENCES

Ahmed, R. S., Ahmed, E. S. A., & Saeed, R. A. (2021). Machine learning in cyber-physical systems in industry 4.0. In *Artificial Intelligence Paradigms for Smart Cyber-Physical Systems* (pp. 20–41): IGI Global.

Akinci D'Antonoli, T. (2020). Ethical considerations for artificial intelligence: An overview of the current radiology landscape. *Diagn Interv Radiol, 26*(5), 504–511. doi:10.5152/dir.2020.19279

Angelopoulos, A., Michailidis, E. T., Nomikos, N., Trakadas, P., Hatziefremidis, A., Voliotis, S., & Zahariadis, T. (2020). Tackling faults in the Industry 4.0 era—a survey of machine-learning solutions and key aspects. *Sensors, 20*(1). doi:10.3390/s20010109

Babiceanu, R. F., & Seker, R. (2016). Big Data and virtualization for manufacturing cyber-physical systems: A survey of the current status and future outlook. *Computers in Industry, 81*, 128–137. doi:10.1016/j.compind.2016.02.004

Bäuerle, A., Cabrera, Á. A., Hohman, F., Maher, M., Koski, D., Suau, X., ... Moritz, D. (2022). *Symphony: Composing interactive interfaces for machine learning. Paper presented at the Proceedings of the 2022 CHI Conference on Human Factors in Computing Systems.*

Blomster, M., & Koivumäki, T. (2022). Exploring the resources, competencies, and capabilities needed for successful machine learning projects in digital marketing. *Information Systems and e-Business Management, 20*(1), 123–169. doi:10.1007/s10257-021-00547-y

Calinescu, R., Cámara, J., & Paterson, C. (2019, May 28–28). *Socio-cyber-physical systems: Models, opportunities, open challenges. Paper presented at the 2019 IEEE/ACM 5th International Workshop on Software Engineering for Smart Cyber-Physical Systems (SEsCPS).*

Carvalho, T. P., Soares, F. A. A. M. N., Vita, R., Francisco, R.d.P., Basto, J. P., & Alcalá, S. G. S. (2019). A systematic literature review of machine learning methods applied to predictive maintenance. *Computers & Industrial Engineering, 137,* 106024. doi:10.1016/j.cie.2019.106024

Cheng, X., Chaw, J. K., Goh, K. M., Ting, T. T., Sahrani, S., Ahmad, M. N., ... Ang, M. C. (2022). Systematic literature review on visual analytics of predictive maintenance in the manufacturing industry. *Sensors, 22*(17), 6321. Retrieved from https://www.mdpi.com/1424-8220/22/17/6321

Costa, F., Lispi, L., Staudacher, A. P., Rossini, M., Kundu, K., & Cifone, F. D. (2019). How to foster sustainable continuous improvement: A cause-effect relations map of lean soft practices. *Operations Research Perspectives, 6,* 100091. doi:10.1016/j. orp.2018.100091

Deist, T. M., Jochems, A., van Soest, J., Nalbantov, G., Oberije, C., Walsh, S., ... Lambin, P. (2017). Infrastructure and distributed learning methodology for privacy-preserving multi-centric rapid learning health care: EuroCAT. *Clinical and Translational Radiation Oncology, 4,* 24–31. doi:10.1016/j.ctro.2016.12.004

Dogan, A., & Birant, D. (2021). Machine learning and data mining in manufacturing. *Expert Systems with Applications, 166,* 114060. doi:10.1016/j.eswa.2020.114060

Ehrlinger, L., Haunschmid, V., Palazzini, D., & Lettner, C. (2019). *A DaQL to monitor data quality in machine learning applications. Paper presented at the Database and Expert Systems Applications,* Cham.

Fahle, S., Prinz, C., & Kuhlenkötter, B. (2020). Systematic review on machine learning (ML) methods for manufacturing processes – Identifying artificial intelligence (AI) methods for field application. *Procedia CIRP, 93,* 413–418. doi:10.1016/j. procir.2020.04.109

Florian, E., Sgarbossa, F., & Zennaro, I. (2021). Machine learning-based predictive maintenance: A cost-oriented model for implementation. *International Journal of Production Economics, 236,* 108114. doi:10.1016/j.ijpe.2021.108114

Gohel, H. A., Upadhyay, H., Lagos, L., Cooper, K., & Sanzetenea, A. (2020). Predictive maintenance architecture development for nuclear infrastructure using machine learning. *Nuclear Engineering and Technology, 52*(7), 1436–1442. doi:10.1016/j.net.2019.12.029

Janssen, M., Brous, P., Estevez, E., Barbosa, L. S., & Janowski, T. (2020). Data governance: Organizing data for trustworthy Artificial Intelligence. *Government Information Quarterly, 37*(3), 101493. doi:10.1016/j.giq.2020.101493

Kang, Z., Catal, C., & Tekinerdogan, B. (2020). Machine learning applications in production lines: A systematic literature review. *Computers & Industrial Engineering, 149,* 106773. doi:10.1016/j.cie.2020.106773

Khaledian, Y., & Miller, B. A. (2020). Selecting appropriate machine learning methods for digital soil mapping. *Applied Mathematical Modelling, 81,* 401–418. doi:10.1016/j.apm.2019.12.016

Khan, M. I. H., Sablani, S. S., Nayak, R., & Gu, Y. (2022). Machine learning-based modeling in food processing applications: State of the art. *Compr Rev Food Sci Food Saf, 21*(2), 1409–1438. doi:10.1111/1541-4337.12912

Kim, D., Kang, P., Cho, S., Lee, H.-J., & Doh, S. (2012). Machine learning-based novelty detection for faulty wafer detection in semiconductor manufacturing. *Expert Systems with Applications, 39*(4), 4075–4083. doi:10.1016/j.eswa.2011. 09.088

Kim, S. H., Kim, C. Y., Seol, D. H., Choi, J. E., & Hong, S. J. (2022). Machine learning-based process-level fault detection and part-level fault classification in semiconductor etch equipment. *IEEE Transactions on Semiconductor Manufacturing*, *35*(2), 174–185. doi:10.1109/TSM.2022.3161512

Kourouklidis, P., Kolovos, D., Matragkas, N., & Noppen, J. (2020). *Towards a low-code solution for monitoring machine learning model performance. Paper presented at the Proceedings of the 23rd ACM/IEEE International Conference on Model Driven Engineering Languages and Systems: Companion Proceedings*, Virtual Event, Canada. doi:10.1145/3417990.3420196

Li, G., Zhou, X., & Cao, L. (2021). *Machine learning for databases. Paper presented at the The First International Conference on AI-ML-Systems*, Bangalore, India. doi:10.1145/3486001.3486248

Li, M., & Li, S. (2022). Research on optimization of food industry processing process based on computational intelligence. *Wireless Communications and Mobile Computing*, *2022*, 7781369. doi:10.1155/2022/7781369

Luo, G. (2016). A review of automatic selection methods for machine learning algorithms and hyper-parameter values. *Network Modeling Analysis in Health Informatics and Bioinformatics*, *5*(1), 18. doi:10.1007/s13721-016-0125-6

Ma, S., Zhang, Y., Lv, J., Yang, H., & Wu, J. (2019). Energy-cyber-physical system enabled management for energy-intensive manufacturing industries. *Journal of Cleaner Production*, *226*, 892–903. doi:10.1016/j.jclepro.2019.04.134

Michelson, K. N., Klugman, C. M., Kho, A. N., & Gerke, S. (2022). Ethical considerations related to using machine learning-based prediction of mortality in the pediatric intensive care unit. *The Journal of Pediatrics*, *247*, 125–128.

Mitri, D. D., Scheffel, M., Drachsler, H., Börner, D., Ternier, S., & Specht, M. (2017). *Learning pulse: A machine learning approach for predicting performance in self-regulated learning using multimodal data. Paper presented at the Proceedings of the Seventh International Learning Analytics & Knowledge Conference*, Vancouver, British Columbia, Canada. doi:10.1145/3027385.3027447

Morella, P., Lambán, M. P., Royo, J., Sánchez, J. C., & Ng Corrales, L.d.C. (2020). Development of a new green indicator and its implementation in a cyber–physical system for a green supply chain. *Sustainability (Switzerland)*, *12*(20), 8629. Retrieved from https://www.mdpi.com/2071-1050/12/20/8629

Nagarajan, S. M., Deverajan, G. G., Bashir, A. K., Mahapatra, R. P., & Al-Numay, M. S. (2022). IADF-CPS: Intelligent anomaly detection framework towards cyber physical systems. *Computer Communications*, *188*, 81–89. doi:10.1016/j.comcom.2022.02.022

Naz, F., Agrawal, R., Kumar, A., Gunasekaran, A., Majumdar, A., & Luthra, S. (2022). Reviewing the applications of artificial intelligence in sustainable supply chains: Exploring research propositions for future directions. *Business Strategy and the Environment*, *31*(5), 2400–2423. doi:10.1002/bse.3034

Oliveira, L. M. C., Dias, R., Rebello, C. M., Martins, M. A. F., Rodrigues, A. E., Ribeiro, A. M., & Nogueira, I. B. R. (2021). Artificial intelligence and cyber-physical systems: A review and perspectives for the future in the chemical industry. *AI*, *2*(3), 429–443. Retrieved from https://www.mdpi.com/2673-2688/2/3/27

Olowononi, F. O., Rawat, D. B., & Liu, C. (2021). Resilient machine learning for networked cyber physical systems: A survey for machine learning security to securing machine learning for CPS. *IEEE Communications Surveys & Tutorials*, *23*(1), 524–552. doi:10.1109/COMST.2020.3036778

Qin, S. J., & Chiang, L. H. (2019). Advances and opportunities in machine learning for process data analytics. *Computers & Chemical Engineering*, *126*, 465–473. doi:10.1016/j.compchemeng.2019.04.003

Radanliev, P., De Roure, D., Van Kleek, M., Santos, O., & Ani, U. (2021). Artificial intelligence in cyber physical systems. *AI & Society, 36*(3), 783–796. doi:10.1007/s00146-020-01049-0

Rahmattalabi, A., & Xiang, A. (2022). Promises and challenges of causality for ethical machine learning. *arXiv preprint arXiv:2201.10683.*

Rai, R., Tiwari, M. K., Ivanov, D., & Dolgui, A. (2021). Machine learning in manufacturing and industry 4.0 applications. *International Journal of Production Research, 59*(16), 4773–4778. doi:10.1080/00207543.2021.1956675

Rani, S., Kataria, A., Chauhan, M., Rattan, P., Kumar, R., & Sivaraman, A. K. (2022). Security and privacy challenges in the deployment of cyber-physical systems in smart city applications: State-of-art work. *Materials Today: Proceedings, 62*(7), 4671–4677. doi:10.1016/j.matpr.2022.03.123

Ranjan, N., Kumar, R., Kumar, R., Kaur, R., & Singh, S. (2022). Investigation of fused filament fabrication-based manufacturing of ABS-Al composite structures: Prediction by machine learning and optimization. *Journal of Materials Engineering and Performance.* doi:10.1007/s11665-022-07431-x

Ray, S. (2019, February 14–16). *A quick review of machine learning algorithms. Paper presented at the 2019 International Conference on Machine Learning, Big Data, Cloud and Parallel Computing (COMITCon).*

Rush, B., Celi, L. A., & Stone, D. J. (2019). Applying machine learning to continuously monitored physiological data. *Journal of Clinical Monitoring and Computing, 33*(5), 887–893. doi:10.1007/s10877-018-0219-z

Russo, A. (2021). Some ethical issues in the review process of machine learning conferences. *arXiv preprint arXiv:2106.00810.*

Shaikh, A., Shinde, S., Rondhe, M., & Chinchanikar, S. (2023). Machine learning techniques for smart manufacturing: A comprehensive review. In *2nd International Conference on Industry 4.0 and Advanced Manufacturing, I-4AM 2022* (pp. 127–137): Springer Science and Business Media Deutschland GmbH.

Shin, P. W., Lee, J., & Hwang, S. H. (2020, February. 19–21). *Data governance on business/data dictionary using machine learning and statistics. Paper presented at the 2020 International Conference on Artificial Intelligence in Information and Communication (ICAIIC).*

Shishvan, O. R., Zois, D. S., & Soyata, T. (2018). Machine intelligence in healthcare and medical cyber physical systems: A survey. *IEEE Access, 6,* 46419–46494. doi:10.1109/ACCESS.2018.2866049

Terranova, N., Venkatakrishnan, K., & Benincosa, L. J. (2021). Application of machine learning in translational medicine: Current status and future opportunities. *The AAPS Journal, 23*(4), 74. doi:10.1208/s12248-021-00593-x

Theissler, A., Pérez-Velázquez, J., Kettelgerdes, M., & Elger, G. (2021). Predictive maintenance enabled by machine learning: Use cases and challenges in the automotive industry. *Reliability Engineering & System Safety, 215,* 107864. doi:10.1016/j.ress.2021.107864

Tirkolaee, E. B., Sadeghi, S., Mooseloo, F. M., Vandchali, H. R., & Aeini, S. (2021). Application of machine learning in supply chain management: A comprehensive overview of the main areas. *Mathematical Problems in Engineering, 2021,* 1476043. doi:10.1155/2021/1476043

Wickramasinghe, C. S., Marino, D. L., Amarasinghe, K., & Manic, M. (2018, October 21–23). *Generalization of deep learning for cyber-physical system security: A survey. Paper presented at the IECON 2018 - 44th Annual Conference of the IEEE Industrial Electronics Society.*

Zhang, C., Wang, Z., Ding, K., Chan, F. T., & Ji, W. (2020). An energy-aware cyber physical system for energy Big data analysis and recessive production anomalies detection in discrete manufacturing workshops. *International Journal of Production Research*, *58*(23), 7059–7077.

Chapter 5

Environmental impact of operations and supply chain from Fourth Industrial Revolution and machine learning approaches

Abhishek Bhattacharjee
Lovely Professional University, Phagwara, India

Raman Kumar
Chandigarh University, Mohali, India

5.1 INTRODUCTION

The Industrial Revolution, which started in the late 18th century and lasted through the 19th, was a significant change. The transition from an agrarian-based economy to an industrialized one brought about a wave of technological innovations and transformations in social and economic structures (Shahroom et al., 2018).

The Fourth Industrial Revolution and machine learning approaches have the potential to impact the environment, both positively and negatively significantly. On the positive side, these technologies can increase operations and supply chain management efficiency, reducing waste and energy consumption. For example, predictive maintenance using machine learning can help prevent equipment failures, reducing the need for replacements and minimizing waste. Similarly, optimizing supply chain logistics can reduce transportation emissions and decrease overall environmental impact. However, there are also potential negative environmental impacts (Ghadge & Moradlou, 2020). The increased use of technology and data centers required for machine learning algorithms can lead to higher energy consumption and associated greenhouse gas emissions. Additionally, producing electronic devices and hardware necessary for implementing these technologies can generate significant amounts of electronic waste. Data privacy and security concerns can also arise, as well as potential ethical concerns related to using algorithms and automated decision-making (Ol et al., 1900).

Before the Industrial Revolution, most goods were produced by hand using simple tools and equipment. Introducing new machines and production methods changed how goods were made, leading to significant increases in productivity and efficiency. New power sources such as steam engines, water wheels, and coal allowed for more efficient production, and the creation of new tools such as the spinning jenny and power loom

DOI: 10.1201/9781003453567-5

revolutionized textile manufacturing. The growth of industry and the availability of new factory jobs attracted people from rural areas to cities, leading to significant population growth. As people moved from farms to factories, they also became consumers of new goods produced by the factories, leading to a further increase in production and economic growth (Bai et al., 2020).

The Industrial Revolution significantly changed the world, and those changes are still being felt today. It transformed how goods are produced and distributed, leading to the rise of capitalism and the emergence of the modern global economy. The Industrial Revolution also brought about significant changes in social and political structures, including the rise of the middle class and the labor movement (Luiz et al., 2019).

First Industrial Revolution	The development and implementation of new power sources, such as steam engines and water wheels, as well as the invention of new machines, significantly increased the speed and efficiency of manufacturing. This allowed for the mass production of goods previously made by hand, which led to significant advancements in various industries such as textile, iron, and steel. With machines, the production process became faster, cheaper, and more consistent, increasing productivity and profitability. This laid the foundation for modern mass manufacturing, which relies heavily on machines and automation to efficiently produce large quantities of goods (Moktadir et al., 2018).
Second Industrial Revolution	The tech revolution, or the Second Industrial Revolution, occurred between the late 19th and early 20th centuries. A wave of new technological advancements characterized it, including the development of assembly lines and the widespread use of electricity, oil, and gas as power sources. The assembly line, pioneered by Henry Ford, greatly increased efficiency and productivity by dividing the manufacturing process into smaller, specialized tasks that workers could complete quickly and efficiently. This allowed for the mass production of goods much faster than ever before. The widespread use of electricity, oil, and gas also allowed for new machinery and equipment creation, which further automated the manufacturing process. As a result, new industries were developed, old ones were expanded, and new employment opportunities for those with the necessary skills to operate and maintain the new machinery were created. Improved communication technology, such as the telephone and telegraph, allowed for faster and more efficient communication between businesses, suppliers, and customers, further accelerating the pace of industrialization and the growth of the global economy. Overall, the Second Industrial Revolution brought about significant advancements in technology, production methods, and communication, laying the foundation for the modern industrialized world we live in today (Oscar et al., 2021).

Third Industrial Revolution	The digital revolution, often known as the Third Industrial Revolution, started in the middle of the 20th century and was characterized by the widespread use of computers, telecommunications, and data analysis in manufacturing processes. The introduction of computers and data analysis allowed for automating many processes previously done manually or by using simple machines. Programmable logic controllers (PLCs) were used to automate industrial processes and gather data, which helped companies optimize production, reduce waste, and improve quality. The development of modern telecommunications, including the internet, enabled companies to communicate and share data in real time, enabling more efficient supply chain management and collaboration between businesses. The digital revolution also saw the development of new technologies, such as robotics and 3D printing, which allowed for even greater automation and customization of manufacturing processes. These advancements led to significant improvements in production efficiency and product quality, as well as new industries and job opportunities (Health, 2019).
Fourth Industrial Revolution	The present wave of technological breakthroughs in manufacturing and other industries is called the Fourth Industrial Revolution, or Industry 4.0. It is characterized by integrating smart machines, the Internet of Things (IoT), big data, and artificial intelligence (AI) into the production process, aiming to make it more efficient, flexible, and responsive to customer needs. The use of smart machines and factories enables real-time monitoring of production processes, which allows for better decision-making and optimization of production flows. This, in turn, increases efficiency, reduces waste, and improves product quality. Integrating big data analytics and AI enables better predictive maintenance of equipment and tools and facilitates mass customization, where products can be produced in small batches according to individual customer preferences. The Internet of Things (IoT) also enables better communication and collaboration between machines, tools, and people, enabling a highly integrated and automated manufacturing process (Lee et al., 2018).

By merging cutting-edge technologies like the IoT, cloud computing, analytics, artificial AI, and machine learning, Industry 4.0 is revolutionizing the manufacturing sector. The IoT is being used to connect machines, tools, and other devices in the factory, enabling them to communicate with each other and share data in real time. This connectivity enables real-time monitoring of production processes, enabling manufacturers to identify and resolve issues quickly and improve production efficiency. Cloud computing allows manufacturers to store and access data and applications on remote servers, reducing the need for on-site hardware and enabling remote access to production data. Analytics is used to extract insights from data collected from various sources within the factory, including machines, sensors, and other devices. These insights can help manufacturers to identify inefficiencies and opportunities for optimization. AI and machine learning are used to automate and optimize manufacturing processes, from predictive maintenance to supply chain management (Trakadas et al., n.d.). AI-powered systems can analyze vast amounts of data and make real-time decisions, enabling manufacturers to improve quality, reduce waste, and increase efficiency. Industrial automation is using advanced sensors, embedded software, and robotics in manufacturing and other industrial settings to automate and optimize production processes. These technologies collect and analyze data from various sources, including sensors, machines, and production lines, to improve efficiency, quality, and safety. By integrating operational data from ERP (enterprise resource planning), customer service, supply chain, and other company systems Manufacturers can acquire useful insights into their operations and make better decisions with data from production processes. For example, by analyzing data from production lines, manufacturers can identify bottlenecks and inefficiencies, optimize production schedules, and reduce downtime. By linking production data with customer service data, manufacturers can gain insights into customer preferences and improve product design and quality. The use of digital technology in manufacturing, also known as Industry 4.0, can revolutionize the sector by increasing automation, improving efficiency, and enabling real-time data analysis for better decision-making.

Smart factories, which use advanced technologies like IoT sensors, AI, and cloud computing to connect machines, processes, and people, can help manufacturers optimize their production processes, reduce downtime, and improve product quality. With real-time visibility into manufacturing assets and processes, manufacturers can quickly identify and address issues, predict and prevent equipment failures, and improve overall productivity. The benefits of Industry 4.0 are not limited to any specific industrial sector. All industrial businesses can benefit from Industry 4.0 technologies and concepts to streamline operations, cut costs, and increase customer satisfaction. These businesses include discrete and process manufacturing, oil and gas, mining, and other industrial divisions. By embracing digital transformation, industrial companies can stay competitive in an ever-changing business environment and meet the demands of today's digital-savvy customers (Rai et al., 2021).

AI and machine learning have the potential to transform the manufacturing industry by enabling companies to collect, analyze, and act on vast amounts of data generated by their operations, business units, partners, and external sources. Manufacturing firms can gain valuable insights into their production processes and business operations by using AI and machine learning algorithms. They can identify patterns and anomalies in the data, predict and prevent equipment failures, optimize production schedules, and automate decision-making processes. Predictive maintenance is one of the most promising applications of AI and machine learning in manufacturing.

By analyzing data from sensors and other sources, manufacturers can identify potential issues before they occur and take corrective action to avoid downtime and reduce maintenance costs. Predictive maintenance can also improve the overall reliability and performance of industrial equipment and increase uptime and efficiency (Tjahjono, 2017).

A smart factory is an advanced manufacturing facility that leverages advanced technologies like AI, machine learning, and IoT to create an interconnected network of machines, communication mechanisms, and processing capacity. A smart factory is a cyber-physical system (CPS) that combines physical machines and systems with digital technologies and intelligent software to create a highly automated and connected manufacturing environment. This allows for real-time monitoring of production processes and data analysis to identify patterns and trends that can help optimize production and reduce downtime. Smart factories use various advanced technologies, such as robotics, cloud computing, big data analytics, and cybersecurity, to streamline production and create a more efficient and agile manufacturing process. By leveraging AI and machine learning algorithms, smart factories can learn and adapt to changing conditions, optimizing their operations in real time.

A smart factory is an advanced manufacturing facility that uses cutting-edge technologies to optimize production processes, increase efficiency, and enhance quality. Some of the key characteristics of a smart factory include the following:

- Connectivity: A smart factory is a highly connected environment that integrates machines, sensors, and other devices to create a network of interconnected systems that communicate and share data in real time.
- Automation: Smart factories use advanced robotics and other automated systems to streamline production processes, reduce errors, and increase efficiency.
- Data analytics: Smart factories collect and analyze large amounts of data in real time, using advanced analytics tools to identify patterns and trends, optimize production processes, and reduce downtime.
- AI: Smart factories leverage AI and machine learning algorithms to improve decision-making and optimize production processes, increasing efficiency and reducing cost.
- Flexibility: Smart factories are designed to be highly flexible and adaptable, enabling them to respond quickly to changes in demand, product design, or production processes.
- Sustainability: Smart factories are designed to be environmentally friendly, using energy-efficient systems and sustainable manufacturing practices to minimize their environmental impact.
- Analysis of data for the best decision-making: Manufacturing organizations generate vast amounts of data through embedded sensors and networked equipment, and this data can be used to gain insights

into historical trends, spot patterns, and improve decision-making. By leveraging data analytics, manufacturers can optimize production processes, reduce downtime, and improve quality, among other benefits.

- Integration of information technology and operational technology: Interconnectivity is a key feature of smart factories. In a smart factory, machines, devices, and sensors are connected through a network architecture, allowing them to share real-time data. This interconnectivity enables additional factory assets to access and use real-time data from other machines, devices, and sensors, allowing for a more integrated and efficient production process.

- Tailored manufacturing: One of the key goals of Industry 4.0 is to enable manufacturers to more economically create products tailored to specific client's needs, even in small batch sizes. This is sometimes called "mass customization," which is the ability to produce customized products at scale rather than mass production, which produces identical products in large quantities. Manufacturers may swiftly create small batches of one-of-a-kind products for particular customers by utilizing simulation software programs, novel materials, and technologies like 3D printing. This allows manufacturers to offer their customers more personalized products and services while maintaining the cost efficiencies of mass production.

- Supply chain: Industry 4.0 emphasizes the importance of connecting production operations with the supply chain to optimize the manufacturing process (Fatorachian & Kazemi, 2020). By giving suppliers access to production data, manufacturers can better plan the delivery of raw materials and adjust deliveries if production is delayed or interrupted. By examining elements like weather, transportation, and retailer data, producers may also utilize predictive shipping to dispatch finished items at the ideal time to fulfill consumer demand. This can help reduce inventory costs and improve customer satisfaction by ensuring that products are available when and where they are needed. Another technology that is becoming increasingly important for enabling supply chain transparency is blockchain. It can help improve supply chain visibility and traceability, allowing manufacturers to track the movement of goods and ensure compliance with regulations and standards. Manufacturers can improve supply chain security by using blockchain to create a decentralized ledger of transactions and reduce the risk of fraud and counterfeiting (Hahn, 2019).

Industrialization is necessary for production to be completed efficiently, and businesses can benefit greatly from incorporating information technology (IT). With the advent of Industry 4.0, businesses can now achieve greater quality and customization in their products, leading to improved customer satisfaction and increased competitiveness. However, it is also important to recognize that Industry 4.0 can have significant environmental impacts, including increased consumption of resources, raw materials, energy, and waste and pollution. As a result, businesses and society need to be aware of these dangers and take action to reduce them as it becomes more and more crucial. Using circular economy concepts, which seek to reduce waste and encourage the reuse and recycling of materials, is one strategy to create more sustainable industrialization. Businesses can also deploy energy-efficient technologies and renewable energy sources to lower their energy usage and carbon impact. In order to understand their concerns and work cooperatively to address environmental difficulties, businesses must connect with stakeholders, such as clients, staff, and local communities. Governments and public sector organizations can also play a role by setting regulations and standards to encourage sustainable practices and support businesses to adopt more sustainable technologies and practices.

Industry 4.0 has strongly emphasized production and maximizing profits, which can negatively impact other aspects of society, such as the environment, wealth distribution, and working conditions. It is important to recognize that these issues are interconnected and that focusing solely on profit can lead to unsustainable consumption patterns that harm the environment and society. To address these issues, a shift toward a "strong sustainability" paradigm is necessary, prioritizing the preservation and regeneration of natural resources and the well-being of society over short-term profit maximization. This may be done by implementing the concepts of the circular economy, which minimize waste, encourage material reuse and recycling, and use renewable energy sources and energy-saving devices. In addition, it is important to address issues of wealth distribution and working conditions, as these are also crucial components of sustainable development. This can be achieved by adopting fair labor practices, such as providing fair wages, safe working conditions, and opportunities for skill development and career advancement. There is often a gap between sustainable consumption awareness and actual behavior, and the current focus on production and profit can lead to unsustainable resource use and environmental damage. It is important to recognize that Industry 4.0 technologies can improve sustainability and product quality, but the primary focus is still on boosting output and competitiveness. To address these challenges, it is necessary to prioritize sustainable development and ensure that Industry 4.0 is implemented in a way that considers environmental considerations. This requires a shift in mindset and a willingness to prioritize long-term sustainability over short-term profit. One approach is adopting circular economy concepts, which seek to reduce waste and encourage material reuse and recycling. Businesses can also deploy energy-efficient technologies and renewable energy sources to lower their energy usage and carbon impact. Governments and public sector organizations can also play a role by setting regulations and standards to encourage sustainable practices and support businesses to adopt more sustainable technologies and practices. While there are challenges to implementing sustainable Industry 4.0 practices, such as harmonizing regulations, adopting compatible legal frameworks, and finding skilled workers, achieving sustainable development through collaboration and innovation is possible. By prioritizing sustainability and recognizing the importance of environmental considerations, we can ensure that Industry 4.0 is implemented in a way that benefits both businesses and society.

5.2 OBJECTIVE AND BENEFITS OF INDUSTRY 4.0

Industry 4.0 aims to create a smart factory based on CPS. To manage people and equipment in the actual world, the CPS gathers and analyses data on them. It is a theory that seeks to improve the effectiveness and efficiency of social systems. Manufacturing facilities have already started using sensors to collect data and utilizing that data for monitoring and controlling industrial plants. The future growth of sensor networks and sensor analysis will pave the way for creating "smart factories," or autonomous, intelligent factories. In a smart factory, entire pieces of equipment are networked to:

- Reduce production costs
  While unattended operations have the potential to reduce labor costs and increase efficiency, they also pose significant risks if not properly managed. Failures and abnormalities can lead to downtime and decreased availability of production facilities, resulting in lost revenue and increased costs for recovery. To mitigate these risks, it is essential to implement proactive maintenance strategies, such as predictive maintenance, that use real-time data and machine learning algorithms to identify potential failures before they occur. Additionally, having a well-trained and skilled workforce that can respond quickly and effectively to any abnormalities or failures is crucial. Another important consideration is the ethical implications of relying solely on automated systems. As Industry 4.0 progresses, there is a risk of devaluing the importance of human labor and expertise. It is essential to balance utilizing technology to improve efficiency and sustainability while valuing human workers' role in production.
- Increase production quality
  Industry 4.0 utilizes AI and other advanced technologies to enable efficient and high-quality manufacturing processes. AI in quality control enables producers to monitor and analyze production data in real time, instantly identifying and resolving quality concerns. This also ensures that the components and raw materials used in production are consistently high quality, leading to higher-quality finished products. Additionally, Industry 4.0's emphasis on customization and small-batch production allows for greater flexibility in manufacturing, enabling companies to respond quickly to changing customer demands and market trends.

5.3 BENEFITS OF INDUSTRY 4.0 CONCERNING THE MODERN ERA

Compliance with regulations and standards is paramount in regulated industries such as healthcare. Industry 4.0 technologies can provide benefits, such as improved quality control and traceability of products, but these benefits must be balanced with the regulatory requirements for the safety and efficacy of medical devices, drugs, and biopharmaceuticals. It is important to ensure that putting Industrial 4.0 technology into practice in these industries complies with the applicable regulations and standards and that proper validation and documentation are in place to demonstrate the safety and efficacy of the products.

- Enhanced efficiency and reduced machine downtime: Using cutting-edge technologies like IoT, big data analytics, and AI, Industry 4.0 technologies strive to optimize production processes, boost productivity, and decrease waste and downtime. By integrating machines and systems, collecting and analyzing data in real time, and automating decision-making processes, manufacturers can identify inefficiencies and bottlenecks in their operations, make data-driven improvements, and increase throughput while reducing costs. This leads to higher OEE, which measures the efficiency and effectiveness of production equipment and, ultimately, greater profitability for the company.
- Greater effectiveness: Besides the efficiencies mentioned, Industry 4.0 technologies also enable predictive maintenance, allowing for early detection of potential equipment failures and avoiding costly unplanned downtime. This is accomplished by the real-time monitoring of machine performance and health through sensors and analytics. Furthermore, using digital twins, or virtual replicas of physical assets, allows for simulations and testing of various scenarios before implementing changes in the production line, resulting in reduced risk and downtime. Finally, collaborative robots, or cobots, can improve efficiency and productivity by working alongside human operators to perform repetitive or dangerous tasks, freeing human labor for more complex and creative tasks.
- Enhanced safety and distribution network planning: By enabling real-time tracking and monitoring of products as they move through the distribution network, Industry 4.0 technology can significantly improve supply chain visibility. This can aid in locating bottlenecks and other problems, resulting in a more effective and efficient distribution system. Additionally, this increased visibility can help to improve communication and collaboration across the supply chain, leading to better decision-making and faster response times. Ultimately, this can help to improve customer satisfaction and increase profitability.

- Digital twins for an improved product, production line, and factory lifecycles: This is a main component of Industry 4.0 and is increasingly being adopted by manufacturing companies. By creating a virtual replica of a physical asset, such as a machine or a factory, manufacturers can simulate different scenarios, test new designs, and optimize performance without physical experimentation. This helps reduce costs, speed up development, and enables more efficient maintenance and better decision-making based on real-time data.
- Empowered people: Workers can now access real-time data and analytics, thanks to Industry 4.0 technologies, which gives them a greater understanding of the manufacturing process and helps them spot areas that need development. By automating repetitive tasks, workers can focus on more complex and value-adding activities that require human intervention, such as problem-solving, decision-making, and innovation. This improves productivity and efficiency and provides employees with a more fulfilling work experience.
- Increased collaboration and information exchange: Industry 4.0 technologies allow for increased connectivity and communication among different machines, systems, and facilities, which can help break down traditional silos and enable more collaboration and knowledge sharing. This can result in faster problem-solving, improved decision-making, and better performance and productivity. It also allows for better visibility and control over your entire operation, regardless of scale or location.
- Agility and flexibility: Manufacturing facilities may more easily respond to changes in demand and market conditions, thanks to Industry 4.0 technology, which enables more agile production processes. As you mentioned, this can include scaling up or down, but it can also mean quickly reconfiguring production lines to produce new products or customizing products for individual customers. This increased agility allows manufacturers to respond more quickly to changing customer needs and market trends, giving them a competitive advantage.

5.4 IMPACT OF SUPPLY CHAIN 4.0

Many impacts affected the supply chain listed:

- Improved visibility and transparency: A digital supply chain enables better tracking and monitoring of goods, inventory, and logistics. This lets companies control their supply chain operations better and make more informed decisions.
- Increased efficiency: Automation and digitization of supply chain processes increase the speed and accuracy of operations, reducing manual errors and improving overall efficiency.

- Enhanced customer experience: Supply chain optimization enables faster delivery of goods, better quality control, and more accurate order fulfillment. This leads to improved customer satisfaction and loyalty.
- Better inventory management: A digital supply chain allows real-time inventory tracking, reducing stockouts, and overstocking and minimizing inventory holding costs.
- Cost savings: By improving visibility, efficiency, and inventory management, a digital supply chain can help companies reduce costs and improve their bottom line.
- Flexibility and agility: A digital supply chain can quickly adapt to changing market conditions, customer demands, and supply chain disruptions, making it more flexible and agile.
- Collaboration and communication: A digital supply chain enables better collaboration and communication among supply chain partners, improving overall supply chain coordination and efficiency.
- Increased reliability and transparency: With digital technologies like IoT, big data analytics, and cloud computing, it is possible to track the entire supply chain in real time, which may greatly enhance inventory management and lower the risk of stockouts. This, in turn, can help businesses optimize their supply chain and reduce associated costs. Real-time asset tracking can also help companies quickly identify and address supply chain disruptions, minimizing the impact on production and revenue.
- Developing decisions based on data to reduce costs: Accurate demand forecasting is key to optimizing inventory levels and reducing costs. With more accurate predictions of future demand, businesses can better align their production and inventory levels, ensuring that they have enough stock to meet demand without holding excess inventory. This can help to reduce inventory carrying costs, minimize waste, and prevent stockouts or overstocking. Using machine learning algorithms and predictive analytics, businesses can also identify patterns and trends in demand that might be missed using traditional forecasting methods. This can help to improve the accuracy of demand predictions and enable businesses to respond more effectively to changes in customer demand.
- Enhanced connectivity and collaboration: Information can flow effortlessly between producers, suppliers, and customers with a fully integrated, digital supply chain management system, improving collaboration and removing silos. The common platform enables ongoing planning and increases support, trust, and shared planning solutions among stakeholders. Collaborative planning and knowledge exchange among non-competitive relationships can lead to cost reduction and better practices. The networked platform can also shorten lead times by enhancing communication, enabling suppliers to raise warnings

early, and empowering enterprises to be more risk-responsive. Demand and supply planning can be combined with pricing decisions through closed-loop planning, allowing companies to alter prices in response to expected demand, stock levels, and replenishment capacity. This boosts sales and optimizes inventory.

- A better warehouse management system: Digitalization can also enable automated warehouse operations, such as autonomous mobile robots that can move and transport goods within the warehouse, reducing the need for human labor and increasing efficiency. Warehouse management systems can use AI and machine learning algorithms to optimize goods placement and route warehouse workers and robots. This improves the speed and accuracy of order fulfillment and minimizes the time goods spend in the warehouse with minimal errors, thereby reducing inventory holding costs

- A "smart" supply chain: Supply chains incorporating machine learning algorithms can "learn" from past experiences and identify potential hazards or disruptions before they occur. By continuously analyzing data and identifying patterns, the algorithms can make recommendations to optimize the supply chain and minimize the impact of potential disruptions. Additionally, these systems can automatically adjust to changing conditions in real time, reducing the need for manual intervention and allowing for greater agility and flexibility in response to unexpected events.

- Enhanced agility: Businesses can improve their agility, speed, and flexibility by utilizing a supply chain cloud and supply chain as a service (SCaaS). With real-time, end-to-end visibility and insights, they can make faster, data-driven decisions and quickly adapt to disruptions or changes in demand. The SCaaS model allows businesses to outsource certain aspects of their supply chain management, such as transportation or warehouse management, to third-party providers. This can lower expenses and increase efficiency. Using innovative supply chain solutions and embracing digitalization are essential for keeping a competitive edge in today's quickly evolving business landscape.

5.5 MACHINE LEARNING

AI's area of machine learning is concerned with using data and methods to allow computers to learn and advance over time without explicit programming. This technology has made many applications possible, including self-driving cars and recommendation engines. It is also a crucial tool for data scientists who employ statistical methods to uncover patterns and make predictions in huge datasets. As the use of big data continues to grow, the demand for data scientists who can harness the power of machine learning is likely to increase.

5.5.1 How machine learning works

With data analysis and pattern recognition, a type of AI known as machine learning, computers may learn and develop, similar to how humans learn from their experiences. This process requires minimal human intervention, and the algorithm can be divided into three primary components.

5.5.1.1 A decision process

Machine learning algorithms can generate predictions or categorize objects based on incoming data – which may be labeled or unlabeled. These algorithms analyze patterns in the data and generate estimates based on that analysis. Making decisions can be a complex process that involves integrating information from multiple sources with varying levels of certainty and considering possibilities at different levels of abstraction.

5.5.1.2 An error function

In machine learning, the accuracy of a model's prediction is measured using an error function, which compares the predicted outcomes with known examples. When making predictions for new data, the algorithm applies the model trained on past data to estimate probable values for an unknown variable. This enables businesses to make informed decisions based on predicted outcomes. By analyzing past data and making predictions about future outcomes, machine learning provides valuable insights that businesses can use to create measurable value.

5.5.1.3 A model optimization process

In machine learning, the model is trained through an iterative optimization process to find the maximum and minimum functions. This desire for better outcomes is one of the most important phenomena in machine learning. In each iteration, hyperparameters are adjusted to achieve better results and compare the outcomes. The goal is to create a reliable model with a lower error rate. The weights are modified to minimize the difference between the known example and the model estimate, resulting in better fitting the data points in the training set. The "evaluate and optimize" process is repeated by the algorithm, with weights updated automatically until a predetermined level of accuracy is achieved.

5.5.2 Machine learning approaches

This powerful data analysis technique allows computers to learn from experience, much like humans and other animals do. Unlike traditional statistical modeling techniques, machine learning algorithms learn directly from data without relying on pre-existing equations as a model. The algorithm automatically adjusts and improves performance as more data becomes available for learning. In order to train neural networks to recognize patterns and make predictions based on data, a subset of machine learning called deep learning is used. In order to build a machine learning model, relevant data about the application is collected and analyzed. This data is typically divided into two sets, one for training the model and one for testing its accuracy and performance.

5.5.2.1 Training data

In machine learning, we use data to train it. The dataset consists of input data (features or predictors) and corresponding output data (labels or target variables). The model's parameters are adjusted once the input data is fed into the algorithm to reduce the discrepancy between the anticipated output and the actual result. Once the model has been trained, it can be used to make predictions based on fresh input data.

5.5.2.2 Validation data

Validation data is a separate set of data used to evaluate the performance of a machine learning model that has been trained using a separate training dataset. The purpose of using validation data is to assess the model's generalization performance on new, unseen data and to prevent overfitting. Overfitting occurs when a model becomes too complex and memorizes the training data, resulting in poor performance when presented with new data.

5.5.2.3 Prevent overfitting

The model is trained on the training data to prevent overfitting, and its performance is evaluated on the validation data. The model is then adjusted or optimized based on the validation data results until the desired performance level is achieved. Once optimized, the model can be evaluated on a separate testing dataset to ensure its generalization performance further.

5.5.2.4 Test data

Although test and validation data have similarities, they are not the same. A final model's performance is assessed using test data after training and validation. Contrarily, throughout the model selection and tuning phase, validation data is used to evaluate several candidate models' performance and guard against overfitting.

To evaluate a machine's effectiveness after training it by looking at how well it predicts the future.

- Overfitting happens when a machine learning model becomes excessively complex and starts to fit the noise in the training data instead of identifying the underlying pattern. Consequently, the model may accurately predict the outputs for the training data but perform poorly on new, unseen data. The model becomes too specialized to the training data and fails to generalize to new data. Overfitting can occur due to various factors, such as having too many parameters relative to the amount of training data, training the model for too many epochs, or having noisy or outlier-containing training data. It is incorrect to suggest that overfitting occurs when a model is biased toward the input and produces inaccurate results for minor variations in input values. Instead, overfitting arises when a model becomes overly complex and cannot generalize to new data, regardless of the degree of variation in the input values.
- Underfitting: Underfitting occurs when a machine learning model is overly simplistic and unable to recognize the underlying patterns in the training data. As a result, both the training data and new, previously unexplored data perform poorly. While inefficient algorithms can contribute to underfitting, they are not the only cause. Underfitting can also happen when a model has too few parameters compared to the complexity of the problem, is not trained for long enough epochs, or is not given enough training data. Factors such as insufficient data or inappropriate feature selection can also lead to underfitting. The only way to overcome underfitting is to experiment with different machine learning methods is not entirely accurate. To address underfitting, one can also try increasing the complexity of the model, adding more features or input data, or adjusting hyperparameters such as learning rate or regularization strength. Diagnosing the specific cause of underfitting is important to determine the appropriate corrective measures.

Creating algorithms that can learn from data and make predictions or judgments without explicit programming is known as machine learning, a branch of AI. To create a model that can be used to predict outcomes on fresh, untainted data, machine learning algorithms need sample data, sometimes referred to as training data. The three main groups of machine learning approaches are as follows.

5.5.3 Supervised machine learning

In supervised machine learning, the algorithm is trained on a labeled dataset, where the inputs and their matching outputs are previously known. A mapping between inputs and outputs must be learned as part of supervised learning for the algorithm to predict the outcome for brand-new, unknown inputs correctly. A feature vector, a collection of numerical or categorical values that describe the input, is often how the input data is represented in supervised learning. However, depending on the issue, the output data is typically displayed as a single number or a collection of values (Mahesh, 2020). Algorithms for supervised learning can be applied to various applications, including classification, regression, and sequence prediction. Decision trees, logistic regression, support vector machines, and neural networks are typical examples of supervised learning algorithms. The labeled dataset is often divided into training and validation sets to train a supervised learning algorithm. The algorithm is tested on the validation set after being tested on the training set. The objective is to reduce the discrepancy between the anticipated and actual output on the validation set to ensure that the method generalizes successfully to new, untested data. The algorithm can be used to make predictions on fresh inputs after training. Problems with supervised machine learning can be divided into two categories, which are listed below:

- Classification
- Regression

5.5.3.1 Classification

Predicting a categorical or discrete value using a set of input features is a machine learning task known as classification. Classification aims to learn a decision boundary that divides several classes in the input space so the algorithm can accurately predict the class of new, unseen inputs. In classification, the input data is typically represented as a feature vector, a set of numerical or categorical values describing the input. The output data is a categorical or discrete label that indicates the class of the information. Classification algorithms can be used for various tasks, such as email spam filtering, image recognition, sentiment analysis, and medical diagnosis. The classification techniques of logistic regression, decision trees, support vector machines, and neural networks are among the examples that are frequently used. A labeled dataset is often divided into training and validation sets to train a classification algorithm. The algorithm is tested on the validation set after being tested on the training set. The discrepancy between the predicted and actual classes on the validation set must be minimal for the algorithm to generalize successfully to new, unknown data. After training, the classification method can be used to anticipate the class of new inputs. The algorithm outputs the projected class label based on the input features. The predicted label can be used for various applications, such as filtering spam emails or diagnosing medical conditions.

A classification problem aims to learn a decision boundary that delineates distinct classes in the input space. The output variable in a classification problem is categorical or discrete. The classification algorithm predicts the categories in the dataset based on the input features. Categorical output variables frequently include options like "Ok" or "Nah," Male or Female, Red or Blue, and so forth. Classification methods can solve many problems, such as spam detection, email filtering, image recognition, sentiment analysis, and medical diagnosis. In spam detection, for example, the goal is to classify incoming emails as either spam or non-spam. The input features might include the sender's email address, subject, and body. The classification algorithm would learn to identify patterns in the input features associated with spam emails and use those patterns to predict whether a new email is a spam.

Similarly, email filtering aims to classify incoming emails into different categories, such as work, personal, and promotional emails. The sender's email address, subject, and body are examples of input characteristics. To forecast the category of a new email, the classification algorithm would train itself to recognize patterns in the input features associated with each category.

The following is the list of well-liked classification algorithms:

- Random forest algorithm
- Decision tree algorithm
- Logistic regression algorithm
- Support vector machine algorithm

5.5.3.2 Regression

A particular kind of machine learning problem called regression entails making predictions about continuous values based on input features. For the algorithm to correctly forecast the output for novel, unforeseen inputs, regression aims to learn a mapping between the inputs and a continuous output variable. As a set of numerical or categorical values that describe the input, a feature vector is commonly used to represent the input data in regression. The output data is a continuous value that can be any real number. Regression algorithms can be used for various tasks, such as predicting stock prices, estimating housing prices, and forecasting sales revenue. Some common examples of regression algorithms include linear regression, polynomial regression, decision trees, and neural networks. A labeled dataset is often divided into a training set and a validation set to train a regression algorithm. The algorithm is tested on the validation set after being tested on the training set. The objective is to reduce the discrepancy between the anticipated output and the actual output on the validation set to ensure that the method generalizes successfully to new, untested data. After training, the regression process can forecast the continuous output value for fresh inputs. After considering the input attributes, the algorithm outputs the anticipated output value (Diez-Olivan et al., 2018). The predicted value can then be used for various applications, such as predicting future stock prices or estimating the value of a house.

When an input–output relationship is linear or roughly represented by a linear function, regression modeling techniques describe the relationship between the input variables and a continuous output variable. By minimizing the discrepancy between the expected and actual output values, one can identify a linear function that best fits the data using linear regression. Many issues can be resolved using regression approaches, including forecasting market patterns, weather conditions, and stock prices. In these cases, the input variables might include factors such as historical prices, weather conditions, and other relevant data, and the output variable is a continuous value that represents the predicted value of the target variable.

The following is the list of popular regression algorithms:

- Simple linear regression algorithm
- Multivariate regression algorithm
- Decision tree algorithm
- Lasso regression

5.5.4 Challenges under supervised learning

Supervised learning is a popular machine learning technique for solving problems such as classification and regression. However, several challenges need to be addressed when using supervised learning, including the following:

- Insufficient or biased data: Supervised learning algorithms require labeled data to learn from, but collecting and labeling data can be time-consuming and expensive (Moktadir et al., 2018). Additionally, the data quality can be affected by biases in the data collection process or incomplete data, leading to inaccurate or incomplete models.
- Overfitting: The training set can easily be overfitted by supervised learning methods, which causes the model to perform well on that set but poorly on brand-new, untried data. This is a common issue when the model is too complicated, or the training set is too tiny.
- Underfitting: The model can potentially be overly simplistic and fail to recognize the underlying patterns in the data, which is known as underfitting the data. This can happen when the model is too simple, or the training set is insufficient.
- Feature selection: The quality of the model heavily relies on selecting the input features. Identifying the most relevant features from a large pool of potential inputs can be challenging, especially when the number of features is high.

- Model interpretability: Finally, understanding how the model makes predictions and identifying the factors that contribute to those predictions can be difficult. This can be a significant challenge, especially in domains where the model's output needs to be explained or justified to stakeholders.

Addressing these challenges requires careful attention to data quality, model selection, and validation. It also requires understanding the strengths and limitations of different supervised learning algorithms and techniques and their implications for different types of applications.

5.5.5 Pros and cons of supervised learning

Supervised learning is a powerful machine learning technique that has several advantages and disadvantages, which are discussed below:

5.5.5.1 Pros

- Effective for classification and prediction tasks: Supervised learning is an effective approach for classification and prediction tasks. The objective is to find patterns in the input data that can reliably predict the output when the output is known.
- Able to generalize: Models for supervised learning can extrapolate from training data to produce precise predictions on fresh, unforeseen data. This makes it a powerful tool for applications where accurate predictions are critical.
- Wide range of algorithms: Numerous algorithms for supervised learning, ranging from straightforward linear regression to intricate neural networks, can be used on various datasets and challenges.
- Fast and scalable: Supervised learning algorithms can be trained on large datasets using parallel computing, which makes them fast and scalable for large-scale applications.
- Easy to evaluate: Standard measures like accuracy, precision, and recall may be used to quickly assess the performance of supervised learning models, making it simple to compare various models and choose the most effective one (Burrell, 2016).

5.5.5.2 Cons

- Dependence on labeled data: Supervised learning models depend on labeled data, which can be expensive and time-consuming to collect and label. Additionally, the quality of the model heavily relies on the quality of the labeled data.
- Overfitting: The training data can be readily overfitted by supervised learning models, which causes them to perform well on that data but

poorly on fresh, untried data. This may be an issue when the model is too complicated, or the training set is too small.

- Limited to known outputs: Supervised learning models are limited to predicting known and labeled outputs, which can be a problem when new or unknown outputs need to be predicted.
- Difficulty in selecting relevant features: Selecting relevant features from the input data can be difficult, especially when dealing with high-dimensional datasets.
- Limited interpretability: When the model output needs to be justified or explained to stakeholders, some supervised learning techniques, such as deep neural networks, might be challenging to interpret.

Distributed and supervisory control systems are widely used in manufacturing industries to improve process efficiencies, reduce downtime, and optimize overall productivity. However, these systems rely heavily on the operator's experience, intuition, and judgment, leading to inconsistencies and errors in decision-making. AI has the potential to address these issues by standardizing and enhancing expert knowledge to create efficient decision-support systems. By leveraging machine learning algorithms and predictive analytics, AI can help identify patterns, anomalies, and inefficiencies in the production process that would be difficult for humans to identify. This has led to an increasing demand for mechanical engineers who are also knowledgeable about AI, as well as other professionals such as process and automation engineers, data scientists, IT and data engineers, and experts in the construction of AI with backgrounds in mechanical and electronics (Villalba-Diez et al., n.d.). These professionals are needed to design, develop, and implement AI-based solutions to optimize manufacturing processes and improve product quality while reducing costs and downtime.

5.5.6 Unsupervised machine learning

Unsupervised machine learning, a subset of machine learning, involves training the algorithm on a dataset without explicit labels or target variables. Unsupervised learning seeks patterns, structures, and correlations within the data without understanding the desired results. The two most popular forms of unsupervised learning are dimensionality reduction and clustering. Without being aware of the groupings beforehand, the algorithm in clustering combines related data points based on their properties. The algorithm in dimensionality reduction lowers the number of variables in the dataset while retaining most of the original data. Several industries use unsupervised learning, including customer segmentation, anomaly detection, image and audio recognition, and natural language processing. It can be constructive when labeled data is hard to come by or expensive to acquire, as it can assist in revealing hidden structures and insights inside complicated datasets. However, unsupervised learning also has some challenges and limitations.

Since there are no target variables to optimize, the algorithm's performance is often more difficult to evaluate objectively. Additionally, the results of unsupervised learning can be more difficult to interpret than supervised learning results, as there may not be clear explanations or insights as to why certain patterns or relationships were discovered (Garay-Rondero et al., 2019).

A good example of how unsupervised learning works is an algorithm that is presented with a dataset of unlabeled images of fruit, and its task is to identify patterns and similarities within the data. It may group images with similar shapes or colors, for example, without prior knowledge of the fruit or its classification. Anomaly detection is another kind of unsupervised learning where the algorithm is trained on a dataset of expected or typical behavior and then used to find outliers or abnormalities in fresh data. This can be helpful in situations like fraud detection, when irregular behavior patterns may point to criminal activities. Unsupervised learning can be an effective technique for revealing hidden patterns and insights in data, but it can also be harder to use than supervised learning because there are no predefined labels or target variables to direct the training process. As a result, it can be more difficult to evaluate the performance of unsupervised learning models and interpret their results (Pablo et al., 2020).

Problems with unsupervised machine learning can be divided into two categories, which are listed below"

- Clustering
- Association

5.5.6.1 Clustering

Clustering is a technique in unsupervised machine learning to group similar data points or observations based on their characteristics or features. Clustering aims to find structure in the data without prior knowledge of the groups or labels. The clustering begins with a dataset of observations or data points, each with features or attributes. The algorithm then groups similar data points into clusters based on a similarity metric. The similarity metric could be based on a distance between data points or another similarity measure between their features. Clustering can be done using a variety of algorithms, such as k-means clustering, hierarchical clustering, and density-based clustering. The algorithm choice depends on the data type used and the research objectives. Each algorithm has strengths and drawbacks of its own. Clustering is used in various applications, including customer segmentation, image and pattern recognition, bioinformatics, and social network analysis (Trakadas et al., n.d.). It can help identify groups of similar individuals or objects and provide insights into the underlying structure of the data.

- K-Means clustering algorithm
- Mean-shift algorithm

- DBSCAN algorithm
- Principal component analysis
- Independent component analysis

5.5.6.2 Association

In unsupervised machine learning, the association technique finds relationships or patterns between variables or items in a dataset. Association analysis aims to identify which variables or items tend to occur together and which are more likely to occur independently. Association analysis is typically applied to transactional data, where each transaction consists of items purchased or used together. The algorithm looks for frequent item sets, which are sets of items that occur together in many transactions. The Apriori technique, which generates candidate item sets iteratively and eliminates those that fall below a minimum support criterion, is the most widely used algorithm for association analysis. Market basket analysis is one of the many uses of association analysis that allows businesses to determine which products are usually bought together and utilize this knowledge to develop targeted promotions or product placement plans. It can also be used in recommendation systems, where the algorithm recommends items or products to users based on their past behavior or preferences.

Popular association rule learning algorithms include the following:

- Apriori algorithm
- Eclat
- Principal component analysis
- Dimensionality reduction
- Singular value decomposition
- FP-growth algorithm.

5.5.7 Challenges under unsupervised learning

Unsupervised learning poses several challenges for machine learning practitioners:

- Lack of labeled data: It cannot be easy to gauge an algorithm's performance or evaluate it against other algorithms because unsupervised learning techniques are taught on unlabeled data.
- Difficulty in selecting appropriate algorithms: With unsupervised learning, choosing an appropriate algorithm for a particular problem can be difficult. Many algorithms are available for unsupervised learning, and the choice of algorithm depends on the type of data and the problem at hand (Burrell, 2016).

- Difficulty in interpreting results: Unsupervised learning algorithms can produce complex and intricate models that are difficult to interpret. Understanding what patterns or clusters have been identified and what they can be is challenging.
- Curse of dimensionality: Unsupervised learning algorithms can struggle to work with high-dimensional data when there are many more characteristics than observations. The search space becomes very large, and finding meaningful patterns or clusters becomes more challenging.
- Overfitting: Unsupervised learning algorithms can also suffer from overfitting when the model fits the training data too closely and is overly complex. For fresh, untested data, this may lead to poor generalization performance.

5.5.8 Pros and cons of unsupervised learning

5.5.8.1 Pros of unsupervised learning

- Flexibility: Unsupervised learning does not require labeled data, making it more flexible than supervised learning. This makes it ideal for situations where the data is not readily available, or the labeling process is too expensive or time-consuming.
- Finding hidden patterns: Unsupervised learning can identify hidden patterns, correlations, and groupings that may not be apparent to humans. This can lead to new insights and discoveries that traditional methods may have overlooked.
- Scalability: Unsupervised learning algorithms can be applied to large datasets without manual labeling, saving time and resources.

5.5.8.2 Cons of unsupervised learning

- Difficult evaluation: Since there are no predefined labels or outputs, evaluating the performance of unsupervised learning models can be challenging. It can be difficult to assess the model's precision or determine whether it performs at its peak.
- Lack of control: Unsupervised learning models can behave unpredictably, and there is little control over the output. This can lead to unexpected or inaccurate results.
- Complex algorithms: Many unsupervised learning algorithms are complex and challenging to understand, requiring advanced mathematical and statistical knowledge to develop and interpret the models.

Unsupervised learning can be beneficial in scenarios lacking labeled data or when labeling data is expensive or time-consuming. It can help discover hidden patterns, identify anomalies, and cluster data points. Several industrial automation use cases, including cybersecurity, asset management, condition monitoring, and performance testing, can benefit from unsupervised learning. However, it can also present some challenges, such as the difficulty in evaluating the performance of unsupervised models and the potential for producing irrelevant or misleading results (Trakadas et al., n.d.).

5.6 SEMI-SUPERVISED LEARNING

Supervised and unsupervised learning techniques are combined in this kind of machine learning. The algorithm is given a dataset containing labeled and unlabeled data in semi-supervised learning. The algorithm first learns from and builds a model from the labeled data, which is then used for the unlabeled data to produce predictions. When labeled data collection is expensive or time-consuming, but unlabeled data is widely accessible, semi-supervised learning is beneficial. Semi-supervised learning can increase the precision of predictions on unlabeled data by using a limited amount of labeled data to train the model. The classification of images is one use case for semi-supervised learning. Even though there may be a lot of unlabeled data for photos available, categorizing each image can take a long time. In this instance, the model can be trained using a small quantity of labeled data before applying it to the remaining unlabeled data to provide predictions.

Popular semi-supervised learning algorithms include the following:

- Inductive semi-supervised learning
- Transductive semi-supervised learning

5.6.1 Semi-supervised learning challenges

- Limited labeled data: The availability of labeled data can be a significant constraint in semi-supervised learning. The quantity and quality of the labeled data significantly impact the model's performance.
- Unbalanced labeled data: In many cases, labeled data may be unbalanced, with one class significantly outnumbering the others. This can result in the model being biased toward the majority class and producing inaccurate predictions for the minority class.
- Incorrect labeling: Labeled data can be incorrectly labeled or mislabeled, leading to suboptimal model performance. In semi-supervised learning, incorrect labeling can be particularly problematic since the model relies on labeled and unlabeled data.
- Difficulty in selecting relevant unlabeled data: Selecting relevant unlabeled data to add to the model can be challenging since knowing which data will be most helpful in improving model performance is difficult.

- Model complexity: Semi-supervised learning models can be more complex than supervised learning models, as labeled and unlabeled data must be considered. Longer training times and more complicated algorithms may result from this.

5.6.2 Pros and cons of semi-supervised learning

5.6.2.1 Pros

- Efficient use of data: Semi-supervised learning is more efficient and economical than fully supervised learning since it can use labeled and unlabeled data.
- Improved performance: Semi-supervised learning can enhance the performance of models over supervised learning, particularly when labeled data is in short supply.
- Scalability: Semi-supervised learning can scale to larger datasets and handle streaming data, which is essential for real-world applications.

5.6.2.2 Cons of semi-supervised learning

- Dependency on unlabeled data: The effectiveness of semi-supervised learning highly depends on the quality and quantity of available unlabeled data. Poor quality or insufficient unlabeled data can negatively impact the model's performance.
- Model complexity: Semi-supervised learning models tend to be more complex than supervised learning models, making them difficult to interpret and explain.
- Limited applicability: Semi-supervised learning may not be suitable for all types of problems, especially those where unlabeled data is not readily available or where labeled data is already abundant.

5.7 REINFORCEMENT MACHINE LEARNING

This machine learning technique is where a task-performing agent learns to interact with the environment by acting and getting feedback through rewards or penalties. The agent must learn an ideal policy linking states to actions to maximize its cumulative reward over time.

5.7.1 Some potential advantages of reinforcement learning

- Ability to learn from experience: Reinforcement learning (RL) agents can learn from past experiences and improve performance. This is particularly useful in dynamic environments where the optimal policy may change over time.

- Flexibility: Robotics, video games, and decision-making are just a few examples of the diverse applications in which RL can be used.
- Ability to handle complex problems: RL can handle complex problems with high-dimensional states and action spaces, which may be difficult or impossible to model using other machine learning techniques.

5.7.2 Some potential challenges and limitations of reinforcement learning

- High computational requirements: RL algorithms can be computationally expensive and require significant training time and resources.
- Lack of interpretability: The decision-making process of an RL agent can be difficult to interpret, which may be a concern in some applications where explaining ability is important.
- Reward engineering: Designing appropriate reward functions for the agent can be challenging and may require domain expertise. Improper reward functions can lead to suboptimal or even harmful behavior.
- Exploration–exploitation tradeoff: RL agents must balance exploring new actions and exploiting actions that have led to high rewards. Finding the right balance can be challenging and affect learning speed and quality.

5.7.2.1 Cons

- Exploration–exploitation dilemma: RL algorithms must balance exploring new actions and exploiting known good ones. Finding this balance is critical to optimizing the reward function and can be a challenge.
- High computational cost: RL algorithms require significant computation, especially when dealing with complex environments.
- Limited interpretability: Recognizing and comprehending how RL models make judgments or select actions can be challenging.
- Delayed rewards: It may be difficult for the algorithm to learn and modify its behavior in some situations since acquiring a reward may take a while.
- Need for continuous learning: RL algorithms must learn continuously as the environment changes or new data becomes available. This can be challenging to implement and manage.
- Overfitting: There is a risk of overfitting in RL, where the algorithm learns to optimize for a specific set of conditions and fails to generalize to new situations.
- Safety concerns: In some applications, RL algorithms may make decisions that have safety implications. Ensuring the safety of these systems is critical, and this can be a challenge in certain domains.

Industrial automation is an area that shows great promise with the application of RL technology. The energy consumption in Google's data centers has been significantly reduced, thanks to RL technology developed by DeepMind. There is a large market for automation solutions, and many startups are developing tools that allow businesses to utilize RL and other methods for industrial applications. Bonsai is one such company. AI is being used to fine-tune machinery and equipment in areas where skilled human operators are currently employed (Luiz et al., 2019).

5.8 CONCLUSIONS

The Fourth Industrial Revolution significantly changed industries' operations and supply chain management. Integrating machine learning approaches has increased the efficiency and effectiveness of the operations and supply chain by providing real-time data analysis, predictive maintenance, and autonomous decision-making capabilities. However, adopting these technologies has also raised concerns about their potential environmental impact. The chapter reviewed the existing literature on the environmental impact of the Fourth Industrial Revolution and machine learning approaches on operations and supply chain management. The findings suggest that these technologies can potentially reduce energy consumption, greenhouse gas emissions, and waste generation by optimizing production processes, reducing transportation costs, and minimizing material waste. However, their impact depends on the way they are implemented and used.

5.8.1 Implications

The industry should adopt sustainable practices and consider the environmental impact of its operations and supply chain. The Fourth Industrial Revolution and machine learning approaches should be directed toward achieving sustainability goals. Policymakers should encourage the adoption of sustainable technologies by providing incentives and regulations. They should also ensure that the potential environmental impacts of these technologies are considered in the policy-making process. The research community should conduct further studies on the environmental impact of the Fourth Industrial Revolution and machine learning approaches in different industries and contexts.

5.8.2 Future scope

The development of sustainability frameworks for adopting the Fourth Industrial Revolution and machine learning approaches in operations and supply chain management. The exploration of the potential of circular economy principles in integrating the Fourth Industrial Revolution and machine learning approaches in the supply chain. The investigation of the impact of the Fourth Industrial Revolution and machine learning approaches on the working conditions of the labor force in the industry. The assessment of the environmental impact of the Fourth Industrial Revolution and machine learning approaches on the supply chain from a life cycle perspective. The examination of the potential of blockchain technology in enhancing the sustainability of supply chains.

REFERENCES

Bai, C., Dallasega, P., Orzes, G., & Sarkis, J. (2020). Industry 4.0 technologies assessment: A sustainability perspective. *International Journal of Production Economics*, 229, 107776. https://doi.org/10.1016/j.ijpe.2020.107776

Burrell, J. (2016). *How the machine 'thinks': Understanding opacity in machine learning algorithms*. June, 1–12. https://doi.org/10.1177/2053951715622512

Diez-Olivan, A., Ser, J. Del, Galar, D., & Sierra, B. (2018). PT US CR. *Information Fusion*. https://doi.org/10.1016/j.inffus.2018.10.005

Fatorachian, H., & Kazemi, H. (2020). The management of operations impact of Industry 4.0 on supply chain performance. *Production Planning & Control*, 1–19. https://doi.org/10.1080/09537287.2020.1712487

Garay-Rondero, C. L., Martinez-Flores, J. L., Smith, N. R., Omar, S., Morales, C., & Aldrette-Malacara, A. (2019). Digital supply chain model in Industry 4.0. https://doi.org/10.1108/JMTM-08-2018-0280

Ghadge, A., & Moradlou, H. (2020). The impact of Industry 4.0 implementation on supply chains. *Journal of Manufacturing Technology Management*, 31(4), 669–686. https://doi.org/10.1108/JMTM-10-2019-0368

Hahn, G. J. (2019). Industry 4.0: a supply chain innovation perspective. *International Journal of Production Research*, 1–17. https://doi.org/10.1080/00207543.2019.1641642

Health, P. (2019). *Optimization of municipal waste collection routing: Impact of Industry 4.0 technologies on environmental awareness and sustainability*. https://doi.org/10.3390/ijerph16040634

Lee, J., Davari, H., Singh, J., & Pandhare, V. (2018). Industrial artificial intelligence for Industry 4.0-based manufacturing systems. *Manufacturing Letters*, 18, 20–23. https://doi.org/10.1016/j.mfglet.2018.09.002

Luiz, D., Nascimento, M., Garza-Reyes, J. A., & Rocha-Lona, L. (2019). Exploring Industry 4.0 technologies to enable circular economy practices in a manufacturing context A business model proposal. *Journal of Manufacturing Technology Management*, 30(3), 607–627. https://doi.org/10.1108/JMTM-03-2018-0071

Mahesh, B. (2020, October). Machine learning algorithms - A review. *International Journal of Science and Research (IJSR)*, 9(1), 381–386. https://doi.org/10.21275/ART20203995

Moktadir, A., Ali, S. M., Kusi-Sarpong, S., & Shaikh, A. A. (2018). Process safety and environmental protection. *Process Safety and Environmental Protection*. https://doi.org/10.1016/j.psep.2018.04.020

Ol, J., Aburumman, N., Popp, J., & Khan, M. A. (1900). Impact of Industry 4.0 on environmental sustainability, 1–21.

Oscar, E., Nara, B., Becker, M., Cristofer, I., Luis, J., Brittes, G., Moraes, L., Lima, A., & Brittes, L. (2021). Expected impact of Industry 4.0 technologies on sustainable development: A study in the context of Brazil's plastic industry. *Sustainable Production and Consumption, 25*, 102–122. https://doi.org/10.1016/j.spc.2020.07.018

Pablo, J., Cadavid, U., Lamouri, S., Grabot, B., Pellerin, R., & Fortin, A. (2020). Machine learning applied in production planning and control: a state - of - the - art in the era of industry 4.0. *Journal of Intelligent Manufacturing, 31*(6), 1531–1558. https://doi.org/10.1007/s10845-019-01531-7

Rai, R., Tiwari, M. K., Ivanov, D., & Dolgui, A. (2021). Machine learning in manufacturing and industry 4.0 applications. *International Journal of Production Research, 59*(16), 4773–4778. https://doi.org/10.1080/00207543.2021.1956675

Shahroom, A. A., Hussin, N., Shahroom, A. A., & Hussin, N. (2018). Industrial revolution 4.0 and education. *International Journal of Academic Research in Business and Social Sciences, 8*(9), 314–319. https://doi.org/10.6007/IJARBSS/v8-i9/4593

Tjahjono, B., Esplugues, C., Ares, E., & Pelaez, G. (2017). What does Industry 4.0 mean to supply chain?. *Procedia Manufacturing, 13*, 1175–1182. https://doi.org/10.1016/j.promfg.2017.09.191

Trakadas, P., Hatziefremidis, A., & Voliotis, S. (2020). Tackling faults in the Industry 4.0 era—A survey of machine-learning solutions and key aspects, *Sensors, 20*(1), 1–34. https://doi.org/10.3390/s20010109

Villalba-Diez, J., Schmidt, D., Gevers, R., Buchwitz, M., & Wellbrock, W. (2019). Deep learning for industrial computer vision quality control in the printing Industry 4.0. *Sensors, 19*(18), 3987. https://doi.org/10.3390/s19183987

Chapter 6

Machine learning for resource optimization in Industry 4.0 eco-system

Raman Kumar
Guru Nanak Dev Engineering College, Ludhiana, India

Amit Verma
Chandigarh University, Mohali, India

6.1 INTRODUCTION

Reinforcement learning is machine learning (ML) that makes decisions based on rewards and punishments. The algorithm in reinforcement learning interacts with its environment, receives input in the form of rewards and penalties, and learns to make decisions that maximize the reward. A reinforcement learning system, for example, can guide a robot through a maze and discover the exit. ML algorithms can scan enormous volumes of data to detect patterns and forecast outcomes, generating insights that may be used to enhance manufacturing processes and increase overall efficiency (Angelopoulos et al., 2020). The manufacturing industry is undergoing substantial change due to technology breakthroughs and the rise of Industry 4.0. The Fourth Industrial Revolution, or Industry 4.0, is defined by incorporating digital technology into manufacturing, resulting in a more connected, automated, and intelligent manufacturing environment (Diez-Olivan, Del Ser, Galar, & Sierra, 2019).

ML, a subset of artificial intelligence (AI) that includes the creation of algorithms that can learn from data and make predictions or judgments without being explicitly programmed, is one of the fundamental technologies driving this transition. ML algorithms can evaluate enormous volumes of data and detect patterns, making them well-suited for various manufacturing-related applications (Lee & Lim, 2021).

Smart factories are essential to Industry 4.0 since they use modern technology to streamline manufacturing processes and increase efficiency. ML is becoming more crucial in smart factories, opening up new avenues for resource management and propelling the next generation of production (Kempegowda & Chaczko, 2018). ML is in its early stages in manufacturing but has enormous potential for increasing productivity, lowering costs, and improving product quality. This chapter introduces ML for resource

DOI: 10.1201/9781003453567-6

optimization in the Industry 4.0 eco-system and its possible applications in smart factories (B. Chen et al., 2018).

This chapter aims to give an overview of the function of ML in resource optimization in the Industry 4.0 eco-system. The primary objective is to investigate ML applications in smart factories and their influence on resource optimization. The chapter thoroughly explains ML in smart factories and its potential for increasing productivity, lowering costs, and improving product quality. It will cover various topics, such as an overview of Industry 4.0 and smart factories, the benefits of ML in smart factories, and specific ML applications for resource optimization, such as predictive maintenance, quality control, process optimization, inventory management, and energy optimization.

This chapter will also look at the problems and limits of ML for resource optimization, as well as the technology's future directions in the Industry 4.0 eco-system. This chapter will present a balanced assessment of ML's promise for resource efficiency while addressing the problems that must be addressed to achieve its full potential. Finally, this chapter aims to offer a review of ML for resource optimization in the Industry 4.0 eco-system and to emphasize this technology's possible benefits and limitations in smart factories.

6.2 OVERVIEW OF INDUSTRY 4.0 AND SMART FACTORIES

The Fourth Industrial Revolution, also known as Industry 4.0, is a concept that refers to the adoption of digital technology into production, resulting in a more connected, automated, and intelligent manufacturing environment. The term "Industry 4.0" was coined in Germany in 2011 and has since spread worldwide to characterize the new industrial age (Barari, de Sales Guerra Tsuzuki, Cohen, & Macchi, 2021). One of the primary technologies driving Industry 4.0 is the Internet of Things (IoT), which integrates digital devices into manufacturing, allowing for real-time data gathering and analysis. Big data, cloud computing, AI, and robots are essential technologies in Industry 4.0 (Zhong, Xu, Klotz, & Newman, 2017). Figure 6.1 depicts the main pillar of Industry 4.0.

Smart factories are essential to Industry 4.0 since they use modern technology to streamline manufacturing processes and increase efficiency. Using data and connections to build a more flexible and adaptable industrial environment distinguishes smart factories (Osterrieder, Budde, & Friedli, 2020). Data from multiple sources, such as equipment, sensors, and people, are gathered in smart factories and utilized to make real-time choices that optimize production processes. Smart factories are intended to be more adaptable and sensitive to changes in demand and output, allowing businesses to adjust to new market conditions and client demands swiftly (Grabowska, 2020). Figure 6.2 shows the prominent architecture of smart factories.

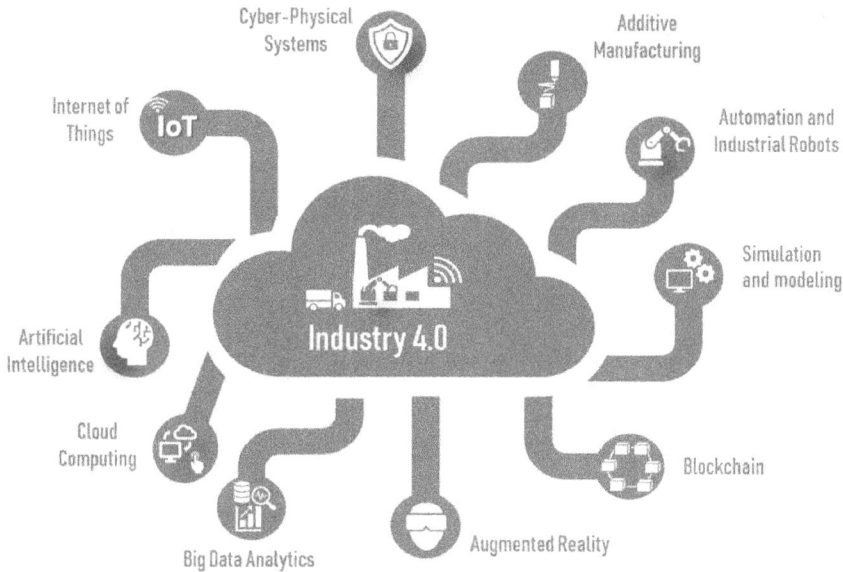

Figure 6.1 Pillars of Industry 4.0. (Adapted from (Ryalat, ElMoaqet, & AlFaouri, 2023) (CC BY 4.0).)

They also enable better monitoring and management of industrial processes, lowering the risk of downtime, and increasing efficiency (Mittal, Khan, Romero, & Wuest, 2018). Industry 4.0 and smart factories represent a significant transition in the manufacturing business, fueled by technical breakthroughs and a shift toward a more connected, automated, and intelligent industrial environment. ML integration is a critical component of this change, opening up new options for resource management and propelling the next generation of production (Shrouf, Ordieres, & Miragliotta, 2014).

6.3 RESOURCE OPTIMIZATION IN SMART FACTORIES

Resource optimization is an essential feature of smart factories and a driving force behind the Industry 4.0 revolution. Resource optimization aims to increase production processes' efficiency and productivity while decreasing waste and maximizing resource use. ML and other modern technologies are used in smart factories to optimize energy, materials, and labor use (Mohamed, Al-Jaroodi, & Lazarova-Molnar, 2019). This optimization is accomplished by gathering and analyzing real-time data, which enables manufacturers to make educated resource utilization decisions and discover areas for improvement. Improved efficiency is one of the primary advantages of resource optimization in smart factories. Smart factories may

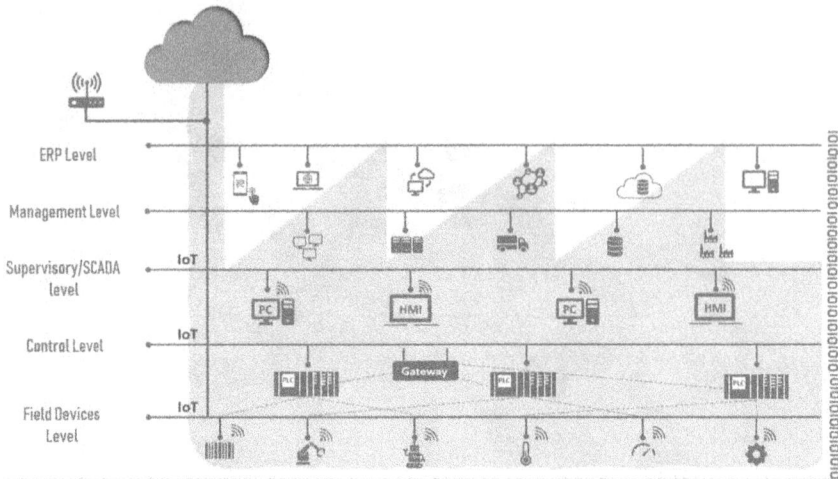

Figure 6.2 Smart factory architecture. (Adapted from (Ryalat et al., 2023) (CC BY 4.0).)

enhance production rates and save costs by maximizing resource utilization and decreasing waste (Afrin, Jin, Rahman, Tian, & Kulkarni, 2019).

Moreover, resource optimization can result in higher-quality products since it enables real-time monitoring of production processes and the early detection of possible problems. Improved sustainability is an advantage of resource efficiency in smart manufacturing. Smart factories may benefit the environment and advance sustainable production techniques by minimizing waste and increasing resource utilization (Wu, Dai, Wang, Xiong, & Guo, 2022). Resource optimization, in summary, is a crucial component of smart factories and a major force behind the Industry 4.0 revolution. ML and other cutting-edge technologies can help smart factories use resources more efficiently while enhancing productivity, product quality, and sustainability (Shi et al., 2020).

6.3.1 Overview of resource optimization

Using resources as effectively and efficiently as feasible is known as resource optimization. By eliminating waste and increasing resource consumption, resource optimization strives to maximize the advantages of resources. Many resources, such as materials, energy, time, and labor, may all be optimized. Resource optimization is crucial for increasing productivity and cutting costs in the industrial sector. Manufacturers may boost output, lower downtime, and enhance product quality by maximizing resource usage (Han et al., 2021).

Predictive maintenance, quality assurance, process optimization, inventory control, and energy optimization are crucial methods for resource optimization in the manufacturing industry. To minimize downtime and increase productivity, predictive maintenance uses ML algorithms to foretell when maintenance will be necessary. ML is used in quality control to track manufacturing processes in real-time and spot possible flaws before they become serious. Process optimization uses ML to streamline industrial procedures while cutting waste and boosting productivity. ML is used in inventory management to optimize inventory levels, reduce waste, and use resources best (Xiong et al., 2020).

ML optimizes manufacturing's energy use while lowering costs and enhancing sustainability. Finally, resource optimization is crucial for increasing productivity, reducing costs, and improving product quality in the industrial sector. Manufacturers may reach unprecedented levels of resource efficiency and power for the next generation of production by utilizing ML and other cutting-edge technology (Khan, Pandey, et al., 2020b).

6.3.2 Applications of machine learning for resource optimization

ML is a potent tool for resource optimization in the industrial sector. ML has several critical applications for resource optimization.

6.3.2.1 Predictive maintenance

Predictive maintenance is an essential application of ML in the industrial sector. Predictive maintenance aims to predict when maintenance is required for equipment to reduce downtime and boost productivity. This approach involves analyzing historical data to identify patterns that indicate when maintenance is required. There are several benefits of using ML for predictive maintenance. First, ML algorithms can analyze vast amounts of data from various sources, such as sensors, maintenance logs, and historical data, to accurately predict when maintenance is required. Second, ML algorithms can monitor equipment health in real-time and alert maintenance teams when issues arise. This approach can help reduce downtime and prevent equipment failure. By predicting when maintenance will need to be performed, ML algorithms may cut downtime and boost productivity (Ong, Wang, Niyato, & Friedrichs, 2022).

6.3.2.2 Quality control

Quality control is another important application of ML in the industrial sector. Quality control aims to identify defects and errors in the manufacturing process and improve product quality. This approach involves continuously monitoring the manufacturing process and identifying possible faults before they become serious. An example of ML for quality control is in the automotive industry, where ML algorithms detect defects in car parts.

In this example, ML algorithms are trained to analyze images of car parts, such as engine components and body panels, to identify defects such as scratches, dents, and cracks. Algorithms for ML may continuously monitor manufacturing processes and spot possible faults before they become serious (Sedighkia & Abdoli, 2022).

6.3.2.3 Process optimization

Process optimization is a critical application of ML in the industrial sector. Process optimization aims to streamline industrial processes, reduce waste, and boost effectiveness. This approach involves identifying areas where processes can be improved and developing new methods to optimize them. An example of process optimization using ML is in the chemical industry, where ML algorithms are used to optimize chemical reactions. In this example, ML algorithms are trained to predict the yield of a chemical reaction based on various process parameters, such as temperature, pressure, and reactant concentration. The trained models can then optimize the process parameters in real time, reducing waste and improving efficiency. Algorithms for ML may be used to streamline industrial processes, cutting waste and boosting effectiveness (Weichert et al., 2019).

6.3.2.4 Inventory management

Inventory management aims to optimize inventory levels, reduce waste, and maximize resource efficiency. This approach involves analyzing historical data and using ML algorithms to predict future demand, enabling businesses to plan their production more effectively and reduce waste. An example of inventory management using ML is in the retail industry, where ML algorithms are used to predict demand for products. ML algorithms are trained to analyze sales data and identify patterns and trends in customer behavior. The trained models can then predict future demand, enabling businesses to optimize their inventory levels and reduce the likelihood of stockouts or overstocking.

Manufacturing companies often have complex supply chains with multiple suppliers, warehouses, and distribution centers. This complexity can lead to inefficiencies, such as overstocking, stockouts, and excess inventory. Manufacturing companies can use ML algorithms to predict demand patterns and optimize inventory levels to address these challenges. ML algorithms are trained to analyze data from various sources, such as sales, supply chain, and historical data. The algorithms can then identify patterns and trends in customer demand, enabling the manufacturing company to optimize its inventory levels and reduce the likelihood of stockouts or overstocking.

ML algorithms can maximize resource efficiency, reduce waste, and optimize inventory levels. ML algorithms may optimize how much energy is utilized throughout manufacturing, cutting costs and enhancing sustainability (Namir, Labriji, & Ben Lahmar, 2022).

6.3.2.5 Predictive scheduling

ML algorithms can also be used for predictive scheduling in various industries. Predictive scheduling involves forecasting future demand patterns and production requirements, allowing businesses to plan their production processes more effectively.

In manufacturing, predictive scheduling can optimize production processes, reduce downtime, and improve product quality. By analyzing data from various sources, such as sales, production, and customer demand, ML algorithms can identify patterns and trends in demand patterns. The algorithms can forecast future demand and production requirements, allowing businesses to plan their production schedules more effectively.

Predictive scheduling can optimize staffing levels and reduce labor costs in the retail industry. By analyzing historical sales and foot traffic data, ML algorithms can predict future demand patterns, allowing businesses to schedule their staff more effectively. This can help to reduce overtime costs and improve customer service levels.

Predictive scheduling can optimize patient care and reduce wait times in the healthcare industry. By analyzing patient data and appointment schedules, ML algorithms can predict future demand patterns and optimize appointment schedules, allowing healthcare providers to provide more efficient and effective patient care. ML algorithms may forecast product demand, enabling businesses to plan their production more effectively and reduce waste (Morariu, Morariu, Răileanu, & Borangiu, 2020).

6.3.2.6 Equipment utilization

ML algorithms can also be used for equipment utilization in various industries. Equipment utilization refers to the effectiveness of equipment used during production processes, and ML algorithms can help improve this efficiency. In the manufacturing industry, equipment utilization can be optimized by analyzing data from various sources, such as equipment logs, production data, and maintenance records. By analyzing this data, ML algorithms can identify patterns in equipment usage, identify bottlenecks, and optimize equipment usage to reduce downtime and improve productivity.

ML algorithms can optimize equipment usage and improve safety in the construction industry. By analyzing data from various sources, such as equipment logs, work schedules, and safety reports, ML algorithms can identify potential safety hazards and optimize equipment usage to minimize these risks. ML algorithms can optimize fleet utilization and reduce fuel costs in the transportation industry. By analyzing data from various sources, such as fuel consumption logs, maintenance records, and driver behavior data, ML algorithms can identify areas of inefficiency and optimize routes and driving behavior to reduce fuel costs and improve overall fleet efficiency.

In the healthcare industry, ML algorithms can be used to optimize the usage of medical equipment and improve patient care. ML algorithms can identify bottlenecks and optimize equipment usage to improve patient care and reduce wait times by analyzing data from various sources such as patient records, equipment logs, and maintenance records. ML algorithms may improve equipment used during production, resulting in less downtime and more productivity (Chen et al., 2019). In the industrial sector, ML is a crucial tool for resource optimization. Manufacturers may optimize resources to new levels using ML algorithms, increasing productivity, cutting costs, and improving product quality (Karkazis, Uzunidis, Trakadas, & Leligou, 2022; Khan, Alsenwi, et al., 2020a; Uzunidis, Karkazis, & Leligou, 2022).

6.3.3 Benefits of machine learning for resource optimization

Resource optimization may benefit greatly from ML in a variety of businesses. These are a few methods ML may improve resource efficiency:

- Predictive maintenance: ML algorithms may assess information from sensors and other sources to forecast when equipment is likely to break down, allowing maintenance to be carried out precisely in time, minimizing downtime, and optimizing resource usage (Ong et al., 2022).
- Demand forecasting: ML may assist firms in accurately predicting customer demand for goods and services, allowing them to allocate their resources better to satisfy that need (Zohdi, Rafiee, Kayvanfar, & Salamiraad, 2022). Planning for production and inventory management can both benefit from this (Nawrocki & Osypanka, 2021).
- Resource allocation: ML algorithms may aid in efficiently distributing resources, including capital, labor, and materials. This may decrease waste, boost productivity, and boost revenue (Huang, van der Aalst, Lu, & Duan, 2011).
- Supply chain optimization: From demand forecasting to controlling inventory levels, transportation paths, and delivery timetables, ML may assist in optimizing supply chain activities. Cost savings and increased customer satisfaction may arise from this (Peng et al., 2019).
- Energy management: Using ML, the energy consumption of factories, buildings, and other facilities may be optimized. To cut expenses and

increase efficiency, algorithms can identify trends in energy use and modify use in real time (Jayaprakash, Nagarajan, Prado, Subramanian, & Divakarachari, 2021).

ML may significantly impact costs, productivity, and profitability regarding resource optimization.

6.4 MACHINE LEARNING ALGORITHMS FOR RESOURCE OPTIMIZATION

The optimization of resources can make use of a variety of ML methods. Some of the most popular algorithms are listed below:

- Linear regression: This straightforward approach may simulate the association between two variables. It may be applied to equipment failure predictions and demand forecasting (Zheng, Ming, & Li, 2017).
- Decision trees: Decision trees may be used to optimize the supply chain and allocate resources. By laying out all potential outcomes and choosing the best one based on the information available, they may assist firms in making decisions (Mastrogiuseppe & Moreno-Bote, 2022).
- Random forest: This decision tree extension combines the forecasts from numerous decision trees that have been trained. It may be applied to resource allocation and demand forecasting (Tyralis, Papacharalampous, & Langousis, 2019).
- Gradient boosting: This sophisticated approach may be used to optimize resource use. It creates a stronger model by combining the predictions of many weak ones (Luo, Yang, Xu, Wang, & Renzo, 2020).
- Support vector machines: These ML methods may be utilized to optimize resource utilization. They have been utilized in various applications for classification and regression issues, including supply chain optimization and energy management (Zendehboudi, Baseer, & Saidur, 2018).
- Neural networks: For complex resource optimization tasks, neural networks are a powerful ML approach. They may be utilized for many things, such as energy management and preventive maintenance (Darshni et al., 2023).

These are only a few illustrations of ML techniques that may be applied to resource optimization. The nature of the data and the project's objectives will determine the ideal method for a specific resource optimization challenge (Weichert et al., 2019).

6.5 CHALLENGES AND LIMITATIONS OF MACHINE LEARNING FOR RESOURCE OPTIMIZATION

6.5.1 Technical challenges

Although there are numerous advantages to ML for resource optimization, there are also certain technical difficulties and restrictions to be aware of (Kumar, Boehm, & Yang, 2017):

- Data quality: ML algorithms depend on high-quality data to provide reliable predictions. The algorithm's predictions might not be correct if the data is noisy, lacking, or inconsistent (Yang, Zhao, & Xing, 2019).
- Feature selection: Selecting the most pertinent variables to the model is known as feature selection. This can be problematic since it's hard to know which characteristics are crucial for a certain issue (Hussain, Hassan, Hussain, & Hossain, 2020).
- Overfitting occurs when an ML model is overly complicated and catches noise in the data rather than the underlying pattern. This can produce subpar performance with fresh data (Karkazis et al., 2022).
- Scalability: Certain ML algorithms may not be able to handle vast volumes of data due to their computational complexity. This can be difficult for resource optimization issues with many variables (Xie et al., 2019).
- Explainability: Some ML algorithms, such as neural networks, can be challenging to understand and explain. Because of this, it may be challenging to discern the algorithm's reasoning process and guarantee the accuracy of the output (Wickramasinghe, Amarasinghe, Marino, Rieger, & Manic, 2021).
- Data privacy: ML algorithms frequently need access to private data, including bank records, operational data, and consumer information. It might be difficult to guarantee the privacy and security of this data (Zelvelder, Westberg, & Främling, 2021).

Thus, even though ML may greatly improve resource optimization, it is crucial to be aware of its technological limitations and difficulties so that they can be handled appropriately.

6.5.2 Data privacy and security concerns

Although ML is an effective technique for resource optimization, several privacy and security issues must be considered. Ensuring private information is not exploited or disclosed is one of the biggest challenges in ML (Xu, Li, Ren, Yang, & Deng, 2019). This can be challenging since the data used to train and test models are frequently gathered from individuals and may contain sensitive personal data that needs to be protected. Large-scale data transmission and storage are also potentially security-risky. Making sure ML models are unbiased and fair is another difficulty. If models are trained with data not representative of the population they are supposed to serve, they may be prejudiced. Discrimination and unequal consequences may result from this (Mohammed, Albeshri, Katib, & Mehmood, 2020). The accuracy and effectiveness of ML models are constrained in addition to these privacy and security issues. These models are prone to errors and adversarial assaults, in which nefarious individuals tamper with the model's inputs to get the desired result. Lastly, it might be challenging to describe ML models' choices. These models are sometimes called "black boxes" since it is challenging to comprehend how they operate and why a particular choice was chosen (Shen, Zhu, Wu, Wang, & Zhou, 2022).

Notwithstanding these difficulties, ML is still an effective technique for allocating resources, and measures may be taken to address privacy and security issues. Sensitive data can be protected using differential privacy and secure multi-party computing techniques. Moreover, utilizing explainable AI strategies can increase model accountability and transparency (M. Li, Yu, Si, Zhang, & Qian, 2022).

6.5.3 Workforce challenges

There are several workforce challenges and limitations of using ML for resource optimization:

- Lack of technical skills: The workforce's capacity to properly apply and use ML algorithms for resource optimization may be constrained by a lack of technical skills. This calls for a sizable investment in training and upskilling to guarantee that the workforce has the required technical skills (Brynjolfsson & Mitchell, 2017).
- Employee reluctance to change: Employees may be reluctant to accept new technology, especially those that automate work. To guarantee that the workforce is on board with ML for resource optimization, it is crucial to convey the advantages and justification for its use properly (Qu, Wang, Govil, & Leckie, 2016).
- Data bias: ML algorithms' accuracy depends on the training data quality. These algorithms may perpetuate and increase biases in resource optimization if the data used to train the algorithms is skewed. Unfair and immoral results may result from this (Tambe, Cappelli, & Yakubovich, 2019).
- Data availability and quality: ML algorithms must be trained on accurate, high-quality data to be effective. Furthermore, it might be difficult to find important data since it could not exist or be hard to obtain (Brynjolfsson & Mitchell, 2017).
- Understanding results: It can be difficult for those without technical skills to understand the output of ML algorithms. This may make it

more difficult to evaluate and track the effects of resource optimization initiatives and come to wise judgments (Tambe et al., 2019).

- Ethical and legal issues: Employing ML to optimize resource usage poses significant ethical and legal issues, including data privacy and the responsible use of algorithms. It is crucial to ensure these issues are addressed and handled publicly to reduce possible harm (Qu et al., 2016).

Resource optimization has the potential to be significantly enhanced by ML. It is critical to be aware of and solve workforce difficulties and constraints to achieve successful adoption and utilization.

6.6 FUTURE DIRECTIONS OF MACHINE LEARNING FOR RESOURCE OPTIMIZATION

Resource optimization is one of the numerous applications of ML, which is a fast-expanding area. Resource allocation, supply chain management, and energy usage reduction are all examples of how to improve the efficiency of resources (Li et al., 2021). There are various areas where ML is anticipated to contribute to resource optimization in the future:

- Predictive maintenance: ML algorithms can foresee when a piece of equipment will break down, enabling businesses to replace or fix it before it causes any interruptions.
- Supply chain optimization: Demand forecasting, inventory control, and transportation planning are just a few supply chain areas that ML can optimize.
- Energy optimization: In many situations, including smart homes and buildings, renewable energy systems, and data centers, ML algorithms may optimize energy use.
- Resource allocation: In complicated systems like distribution networks, transportation networks, and healthcare systems, ML may be utilized to optimize resource allocation.
- Environmental sustainability: ML may assist businesses in choosing more environmentally friendly actions, such as reducing waste production and greenhouse gas emissions.

ML has a promising future in resource optimization, with many intriguing prospects to increase resource use and decrease waste. ML will become crucial as technology advances resource efficiency and environmental initiatives (Y. Chen et al., 2020).

6.7 IMPLICATIONS FOR INDUSTRY 4.0

The widespread adoption of ML in resource optimization has several implications for organizations, including the following:

- More competitiveness: Businesses employing ML to optimize resource usage will probably have an advantage over those not.
- More effectiveness: Organizations may use ML algorithms to optimize resource allocation and cut waste, boosting effectiveness and saving costs.
- Improved sustainability: Utilizing ML to optimize resource consumption may make businesses more environmentally friendly and sustainable.
- New work possibilities: The increasing need for resource optimization ML knowledge will probably lead to new career opportunities across various sectors.
- Technological advancements: The widespread adoption of ML in resource optimization will likely drive technological progress and innovation.
- Ethical and privacy considerations: Organizations must be aware of ML's moral and privacy implications and the potential for algorithm bias.

Ultimately, the widespread use of ML for resource optimization has far-reaching consequences that are expected to influence businesses, industries, and society.

6.8 CONCLUDING REMARKS

By offering fresh and original answers to the problems that businesses encounter, ML has the potential to transform resource efficiency. ML may influence various fields, including environmental sustainability, resource allocation, supply chain efficiency, and predictive maintenance. Organizations may increase resource efficiency, lower waste, and increase sustainability by utilizing the power of ML. ML must be used with other strategies and methodologies to get the best results. This must be kept in mind at all times. Also, enterprises must be mindful of the ethical and privacy concerns of employing these technologies, and ML algorithms must be trained on high-quality data to deliver reliable findings. Firms implementing these technologies expect considerable cost savings, increased efficiency, and better sustainability. ML in resource optimization has a bright future.

REFERENCES

Afrin, M., Jin, J., Rahman, A., Tian, Y.-C., & Kulkarni, A. (2019). Multi-objective resource allocation for Edge Cloud based robotic workflow in smart factory. *Future Generation Computer Systems*, 97, 119–130. doi:10.1016/j.future.2019.02.062

Angelopoulos, A., Michailidis, E. T., Nomikos, N., Trakadas, P., Hatziefremidis, A., Voliotis, S., & Zahariadis, T. (2020). Tackling faults in the Industry 4.0 era—a survey of machine-learning solutions and key aspects. *Sensors*, 20(1). doi:10.3390/s20010109

Barari, A., de Sales Guerra Tsuzuki, M., Cohen, Y., & Macchi, M. (2021). Editorial: Intelligent manufacturing systems towards industry 4.0 era. *Journal of Intelligent Manufacturing*, 32(7), 1793–1796. doi:10.1007/s10845-021-01769-0

Brynjolfsson, E., & Mitchell, T. (2017). What can machine learning do? Workforce implications. *Science*, 358(6370), 1530–1534. doi:10.1126/science.aap8062

Chen, B., Wan, J., Lan, Y., Imran, M., Li, D., & Guizani, N. (2019). Improving cognitive ability of edge intelligent IIoT through machine learning. *IEEE Network*, 33(5), 61–67. doi:10.1109/MNET.001.1800505

Chen, B., Wan, J., Shu, L., Li, P., Mukherjee, M., & Yin, B. (2018). Smart factory of Industry 4.0: Key technologies, application case, and challenges. *IEEE Access*, 6, 6505–6519. doi:10.1109/ACCESS.2017.2783682

Chen, Y., Zheng, B., Zhang, Z., Wang, Q., Shen, C., & Zhang, Q. (2020). Deep learning on mobile and embedded devices: State-of-the-art, challenges, and future directions. *ACM Comput. Surv.*, 53(4), Article 84. doi:10.1145/3398209

Darshni, P., Dhaliwal, B. S., Kumar, R., Balogun, V. A., Singh, S., & Pruncu, C. I. (2023). Artificial neural network based character recognition using SciLab. *Multimedia Tools and Applications*, 82(2), 2517–2538. doi:10.1007/s11042-022-13082-w

Diez-Olivan, A., Del Ser, J., Galar, D., & Sierra, B. (2019). Data fusion and machine learning for industrial prognosis: Trends and perspectives towards Industry 4.0. *Information Fusion*, 50, 92–111. doi:10.1016/j.inffus.2018.10.005

Grabowska, S. (2020). Smart factories in the age of Industry 4.0. *Management Systems in Production Engineering*, 28(2), 90–96. doi:10.2478/mspe-2020-0014

Han, Y., Liu, S., Cong, D., Geng, Z., Fan, J., Gao, J., & Pan, T. (2021). Resource optimization model using novel extreme learning machine with t-distributed stochastic neighbor embedding: Application to complex industrial processes. *Energy*, 225, 120255. doi:10.1016/j.energy.2021.120255

Huang, Z., van der Aalst, W. M. P., Lu, X., & Duan, H. (2011). Reinforcement learning based resource allocation in business process management. *Data & Knowledge Engineering*, 70(1), 127–145. doi:10.1016/j.datak.2010.09.002

Hussain, F., Hassan, S. A., Hussain, R., & Hossain, E. (2020). Machine learning for resource management in cellular and IoT networks: Potentials, current solutions, and open challenges. *IEEE Communications Surveys & Tutorials*, 22(2), 1251–1275. doi:10.1109/COMST.2020.2964534

Jayaprakash, S., Nagarajan, M. D., Prado, R. P., Subramanian, S., & Divakarachari, P. B. (2021). A systematic review of energy management strategies for resource allocation in the cloud: Clustering, optimization and machine learning. *Energies*, 14(17). doi:10.3390/en14175322

Karkazis, P., Uzunidis, D., Trakadas, P., & Leligou, H. C. (2022). Design challenges on machine-learning enabled resource optimization. *IT Professional*, 24(5), 69–74. doi:10.1109/MITP.2022.3194129

Kempegowda, S. M., & Chaczko, Z. (2018, December 18–20). *Industry 4.0 complemented with EA approach: A proposal for digital transformation success*. Paper presented at the 2018 26th International Conference on Systems Engineering (ICSEng).

Khan, L. U., Alsenwi, M., Yaqoob, I., Imran, M., Han, Z., & Hong, C. S. (2020a). Resource optimized federated learning-enabled cognitive internet of things for smart industries. *IEEE Access*, *8*, 168854–168864. doi:10.1109/ACCESS.2020.3023940

Khan, L. U., Pandey, S. R., Tran, N. H., Saad, W., Han, Z., Nguyen, M. N. H., & Hong, C. S. (2020b). Federated learning for edge networks: Resource optimization and incentive mechanism. *IEEE Communications Magazine*, *58*(10), 88–93. doi:10.1109/MCOM.001.1900649

Kumar, A., Boehm, M., & Yang, J. (2017). *Data management in machine learning: Challenges, techniques, and systems*. Paper presented at the Proceedings of the 2017 ACM International Conference on Management of Data, Chicago, Illinois, USA. doi:10.1145/3035918.3054775

Lee, C., & Lim, C. (2021). From technological development to social advance: A review of Industry 4.0 through machine learning. *Technological Forecasting and Social Change*, *167*, 120653. doi:10.1016/j.techfore.2021.120653

Li, M., Yu, F. R., Si, P., Zhang, Y., & Qian, Y. (2022). Intelligent resource optimization for blockchain-enabled IoT in 6G via collective reinforcement learning. *IEEE Network*, *36*(6), 175–182. doi:10.1109/MNET.105.2100516

Li, Y., Lei, G., Bramerdorfer, G., Peng, S., Sun, X., & Zhu, J. (2021). Machine learning for design optimization of electromagnetic devices: Recent developments and future directions. *Applied Sciences*, *11*(4). doi:10.3390/app11041627

Luo, Y., Yang, J., Xu, W., Wang, K., & Renzo, M. D. (2020). Power consumption optimization using gradient boosting aided deep Q-network in C-RANs. *IEEE Access*, *8*, 46811–46823. doi:10.1109/ACCESS.2020.2978935

Mastrogiuseppe, C., & Moreno-Bote, R. (2022). Deep imagination is a close to optimal policy for planning in large decision trees under limited resources. *Scientific Reports*, *12*(1), 10411. doi:10.1038/s41598-022-13862-2

Mittal, S., Khan, M. A., Romero, D., & Wuest, T. (2018). A critical review of smart manufacturing & Industry 4.0 maturity models: Implications for small and medium-sized enterprises (SMEs). *Journal of Manufacturing Systems*, *49*, 194–214. doi:10.1016/j.jmsy.2018.10.005

Mohamed, N., Al-Jaroodi, J., & Lazarova-Molnar, S. (2019). Leveraging the capabilities of Industry 4.0 for improving energy efficiency in smart factories. *IEEE Access*, *7*, 18008–18020. doi:10.1109/ACCESS.2019.2897045

Mohammed, T., Albeshri, A., Katib, I., & Mehmood, R. (2020). UbiPriSEQ—Deep reinforcement learning to manage privacy, security, energy, and QoS in 5G IoT HetNets. *Applied Sciences*, *10*(20). doi:10.3390/app10207120

Morariu, C., Morariu, O., Răileanu, S., & Borangiu, T. (2020). Machine learning for predictive scheduling and resource allocation in large scale manufacturing systems. *Computers in Industry*, *120*, 103244. doi:10.1016/j.compind.2020.103244

Namir, K., Labriji, H., & Ben Lahmar, E. H. (2022). Decision support tool for dynamic inventory management using machine learning, time series and combinatorial optimization. *Procedia Computer Science*, *198*, 423–428. doi:10.1016/j.procs.2021.12.264

Nawrocki, P., & Osypanka, P. (2021). Cloud resource demand prediction using machine learning in the context of QoS parameters. *Journal of Grid Computing, 19*(2), 20. doi:10.1007/s10723-021-09561-3

Ong, K. S. H., Wang, W., Niyato, D., & Friedrichs, T. (2022). Deep-reinforcement-learning-based predictive maintenance model for effective resource management in industrial IoT. *IEEE Internet of Things Journal, 9*(7), 5173–5188. doi:10.1109/JIOT.2021.3109955

Osterrieder, P., Budde, L., & Friedli, T. (2020). The smart factory as a key construct of industry 4.0: A systematic literature review. *International Journal of Production Economics, 221*, 107476. doi:10.1016/j.ijpe.2019.08.011

Peng, Z., Zhang, Y., Feng, Y., Zhang, T., Wu, Z., & Su, H. (2019, November 22–24). *Deep reinforcement learning approach for capacitated supply chain optimization under demand uncertainty. Paper presented at the 2019 Chinese Automation Congress (CAC).*

Qu, S., Wang, J., Govil, S., & Leckie, J. O. (2016). Optimized adaptive scheduling of a manufacturing process system with multi-skill workforce and multiple machine types: An ontology-based, multi-agent reinforcement learning approach. *Procedia CIRP, 57*, 55–60. doi:10.1016/j.procir.2016.11.011

Ryalat, M., ElMoaqet, H., & AlFaouri, M. (2023). Design of a smart factory based on cyber-physical systems and internet of things towards Industry 4.0. *Applied Sciences, 13*(4). doi:10.3390/app13042156

Sedighkia, M., & Abdoli, A. (2022). Linking remote sensing analysis and reservoir operation optimization for improving water quality management of reservoirs. *Journal of Hydrology, 613*, 128445. doi:10.1016/j.jhydrol.2022.128445

Shen, S., Zhu, T., Wu, D., Wang, W., & Zhou, W. (2022). From distributed machine learning to federated learning: In the view of data privacy and security. *Concurrency and Computation: Practice and Experience, 34*(16), e6002. doi:10.1002/cpe.6002

Shi, Z., Xie, Y., Xue, W., Chen, Y., Fu, L., & Xu, X. (2020). Smart factory in Industry 4.0. *Systems Research and Behavioral Science, 37*(4), 607–617. doi:10.1002/sres.2704

Shrouf, F., Ordieres, J., & Miragliotta, G. (2014, December 9–12). *Smart factories in Industry 4.0: A review of the concept and of energy management approached in production based on the Internet of Things paradigm. Paper presented at the 2014 IEEE International Conference on Industrial Engineering and Engineering Management.*

Tambe, P., Cappelli, P., & Yakubovich, V. (2019). Artificial intelligence in human resources management: Challenges and a path forward. *California Management Review, 61*(4), 15–42. doi:10.1177/0008125619867910

Tyralis, H., Papacharalampous, G., & Langousis, A. (2019). A brief review of random forests for water scientists and practitioners and their recent history in water resources. *Water, 11*(5). doi:10.3390/w11050910

Uzunidis, D., Karkazis, P., & Leligou, H. C. (2022, July 20–22). *Machine learning resource optimization enabled by cross layer monitoring. Paper presented at the 2022 13th International Symposium on Communication Systems, Networks and Digital Signal Processing (CSNDSP).*

Weichert, D., Link, P., Stoll, A., Rüping, S., Ihlenfeldt, S., & Wrobel, S. (2019). A review of machine learning for the optimization of production processes. *The International Journal of Advanced Manufacturing Technology, 104*(5), 1889–1902. doi:10.1007/s00170-019-03988-5

Wickramasinghe, C. S., Amarasinghe, K., Marino, D. L., Rieger, C., & Manic, M. (2021). Explainable unsupervised machine learning for cyber-physical systems. *IEEE Access*, *9*, 131824–131843. doi:10.1109/ACCESS.2021.3112397

Wu, Y., Dai, H. N., Wang, H., Xiong, Z., & Guo, S. (2022). A survey of intelligent network slicing management for industrial IoT: Integrated approaches for smart transportation, smart energy, and smart factory. *IEEE Communications Surveys & Tutorials*, *24*(2), 1175–1211. doi:10.1109/COMST.2022.3158270

Xie, J., Yu, F. R., Huang, T., Xie, R., Liu, J., Wang, C., & Liu, Y. (2019). A survey of machine learning techniques applied to Software Defined Networking (SDN): Research issues and challenges. *IEEE Communications Surveys & Tutorials*, *21*(1), 393–430. doi:10.1109/COMST.2018.2866942

Xiong, K., Leng, S., Chen, X., Huang, C., Yuen, C., & Guan, Y. L. (2020). Communication and computing resource optimization for connected autonomous driving. *IEEE Transactions on Vehicular Technology*, *69*(11), 12652–12663. doi:10.1109/TVT.2020.3029109

Xu, G., Li, H., Ren, H., Yang, K., & Deng, R. H. (2019). Data security issues in deep learning: Attacks, countermeasures, and opportunities. *IEEE Communications Magazine*, *57*(11), 116–122. doi:10.1109/MCOM.001.1900091

Yang, J., Zhao, C., & Xing, C. (2019). Big data market optimization pricing model based on data quality. *Complexity*, *2019*, 5964068. doi:10.1155/2019/5964068

Zelvelder, A. E., Westberg, M., & Främling, K. (2021, 2021). *Assessing explainability in reinforcement learning. Paper presented at the Explainable and Transparent AI and Multi-Agent Systems*, Cham.

Zendehboudi, A., Baseer, M. A., & Saidur, R. (2018). Application of support vector machine models for forecasting solar and wind energy resources: A review. *Journal of Cleaner Production*, *199*, 272–285. doi:10.1016/j.jclepro.2018.07.164

Zheng, M., Ming, X., & Li, G. (2017). Dynamic optimization for IPS2 resource allocation based on improved fuzzy multiple linear regression. *Mathematical Problems in Engineering*, *2017*, 2839125. doi:10.1155/2017/2839125

Zhong, R. Y., Xu, X., Klotz, E., & Newman, S. T. (2017). Intelligent manufacturing in the context of Industry 4.0: A review. *Engineering*, *3*(5), 616–630. doi:10.1016/J.ENG.2017.05.015

Zohdi, M., Rafiee, M., Kayvanfar, V., & Salamiraad, A. (2022). Demand forecasting based machine learning algorithms on customer information: An applied approach. *International Journal of Information Technology*, *14*(4), 1937–1947. doi:10.1007/s41870-022-00875-3

Chapter 7

Applications of machine learning in smart factory in fourth-generation industrial environment

Raman Kumar and Chamkaur Jindal
Guru Nanak Dev Engineering College, Ludhiana, India

Raman Kumar
Chandigarh University, Mohali, India

7.1 INTRODUCTION

The Fourth Industrial Revolution, or Industry 4.0, is the name given to the current development in industrial production that emphasizes the incorporation of cutting-edge technologies like the Internet of Things (IoT), artificial intelligence (AI), and machine learning (ML) to boost productivity, quality, and efficiency. The core of Industry 4.0 is smart factories, which use these technologies and give producers a method to meet their output targets (Oztemel & Gursev, 2020). ML, a branch of AI, aims to create algorithms that can continuously improve accuracy by learning from new data. Massive volumes of data may be analysed by ML algorithms, which can then spot patterns and correlations and make predictions based on them. Because of this capability, ML is a potent tool for streamlining production procedures in intelligent factories (Pereira & Romero, 2017).

The first portion of this chapter will give an overview of ML and its many kinds. The chapter is structured as follows. The use of ML in smart factories, including its applications in quality control, predictive maintenance, and production optimization, will be covered in the second section. The third part will discuss the advantages and difficulties of using ML in smart factories. The fourth portion will conclude with a forecast for using ML in the industrial sector. The reader will have a solid knowledge of the uses of ML in smart factories, the advantages and drawbacks of using this technology, and the prospects for ML in Industry 4.0 at the end of this chapter.

7.2 OVERVIEW OF MACHINE LEARNING

The field of AI, known as "machine learning," is concerned with developing algorithms that can learn from data over time and become more accurate without explicit programming (Sarker et al., 2020). It is founded on the idea

DOI: 10.1201/9781003453567-7

that computers can learn from data, identify patterns and correlations, and then make predictions based on these findings. Massive amounts of data are examined by ML algorithms, which then uncover intricate connections and patterns and produce predictions that can guide decision-making (Alanne & Sierla, 2022). Supervised, unsupervised, and reinforcement learning are the three categories of ML.

i. A labelled dataset is used to train the algorithm in supervised learning, a type of ML. After being fed input and output pairs, the algorithm must learn a mapping from the input to the output. For instance, a supervised learning system may be trained on a database of handwritten digit images to identify the digits in recent photos (Khan et al., 2022).

ii. A type of ML called unsupervised learning involves training the algorithm on a dataset without labels. Unsupervised learning looks for structures or patterns in data without labelling. Data points can be grouped using an unsupervised learning method, depending on how similar they are (Alloghani, Al-Jumeily, Mustafina, Hussain, & Aljaaf, 2020).

iii. ML that bases choices on rewards and penalties is known as reinforcement learning. In reinforcement learning, the algorithm interacts with the environment and learns to make decisions that maximize the reward by taking feedback in the form of rewards and penalties. For instance, a robot can be led through a maze by a reinforcement learning system and shown the way out (Mosavi et al., 2020).

7.2.1 Timeline of machine learning in Industry 4.0

In the early 2000s, manufacturers used ML algorithms for quality control and predictive maintenance, which marked the beginning of ML's integration into the manufacturing sector. Unfortunately, ML didn't become a key technology for smart factories until recently, with the emergence of Industry 4.0 (Sony & Naik, 2020).

ML was primarily employed in the early phases of Industry 4.0 to monitor and analyse manufacturing processes. Production machine monitoring, fault detection, and maintenance alert generation were all accomplished using ML algorithms. Because of this, producers could identify and fix problems before they seriously hampered the production process (Ghobakhloo, 2020). Details of the emergence of Industry 4.0 is depicted in Figure 7.1.

ML has become a more important technology in the industrial sector due to technological advancements and the expansion of data availability. Modern uses of ML algorithms include quality assurance, proactive maintenance, and production planning. ML algorithms, for instance, may examine

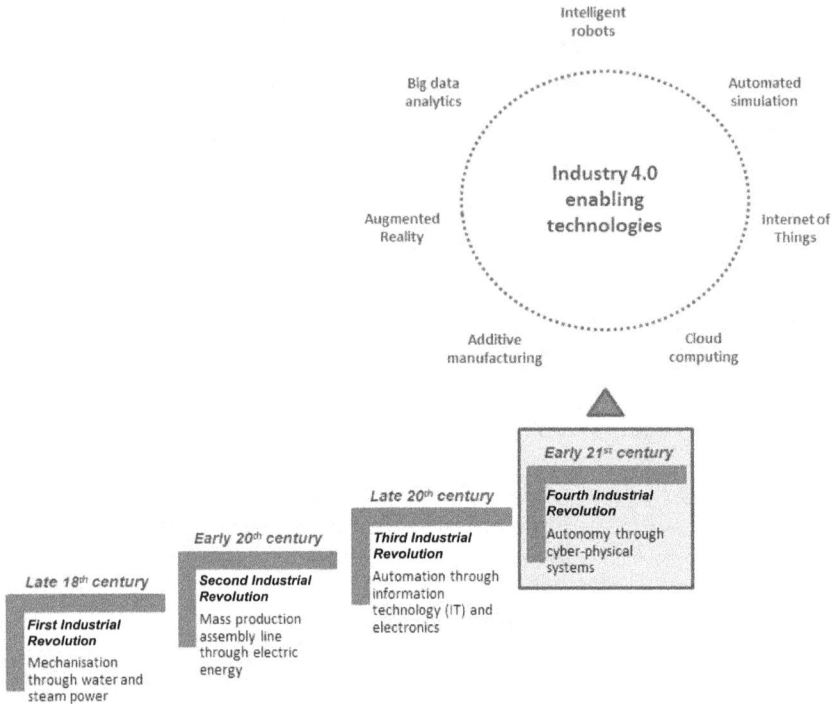

Figure 7.1 Industry 4.0 and its emergence. (Adapted from (Ang, Goh, Saldivar, & Li, 2017) (CC BY 4.0)).

many production data, find patterns and correlations, and forecast things that might help you make decisions (Narciso & Martins, 2020).

Industry 4.0 is anticipated to place even greater emphasis on ML. ML will become a crucial technology for streamlining manufacturing procedures, enhancing quality assurance, and cutting costs due to ML algorithms' rising sophistication and data availability expansion. By enabling manufacturers to react swiftly to shifting market conditions and consumer preferences, ML will also play a crucial role in helping companies satisfy the needs of a changing market (Usuga Cadavid, Lamouri, Grabot, Pellerin, & Fortin, 2020).

Since its early uses in the industrial sector, ML has advanced significantly and is anticipated to play a significant role in Industry 4.0. ML has the potential to revolutionize the manufacturing sector and provide manufacturers with a competitive edge because of its capacity to analyse massive volumes of data, spot patterns and correlations, and create predictions that may guide decision-making (Dalzochio et al., 2020).

7.3 THE USE OF MACHINE LEARNING IN SMART FACTORIES

A manufacturing facility that uses cutting-edge technology and data analytics to streamline production procedures is called a smart factory, also known as an Industry 4.0 factory. A smart factory strives to attain high production levels, adaptability, and efficiency while avoiding waste, risking fewer mistakes, and downtime (Kotsiopoulos, Sarigiannidis, Ioannidis, & Tzovaras, 2021). Smart factories gather, analyse, and act on data produced by industrial processes using a variety of technologies, such as the Internet of Things (IoT), AI, ML, robots, and cloud computing (Malburg, Rieder, Seiger, Klein, & Bergmann, 2021).

This data makes real-time production schedule optimization, equipment performance monitoring, and inefficiency detection possible. Workers' safety, comfort, and well-being are also prioritized in smart factories, which frequently include ergonomic design principles and use technologies like augmented reality (AR) and virtual reality (VR) to give employees better training and visualization tools. In general, smart factories are a big step forward for the manufacturing sector, allowing businesses to become more efficient, competitive, and responsive to shifting consumer needs (Ding et al., 2020). The smart factory framework of Industry 4.0 is shown in Figure 7.2.

The use of ML in smart factories has grown in importance since it gives manufacturers a competitive edge and enables them to keep up with the needs of a market that is changing quickly. This section will examine the applications of ML in smart factories, such as quality assurance, predictive maintenance, and production planning (Park, Li, & Hong, 2020).

7.3.1 Quality control

In order to ensure that goods satisfy the specified standards and specifications, quality control is crucial in the manufacturing process. Automating quality control procedures using ML algorithms lowers the possibility of human mistakes and boosts productivity. Image analysis is one of the most often used uses of ML in quality assurance (Park et al., 2020). ML

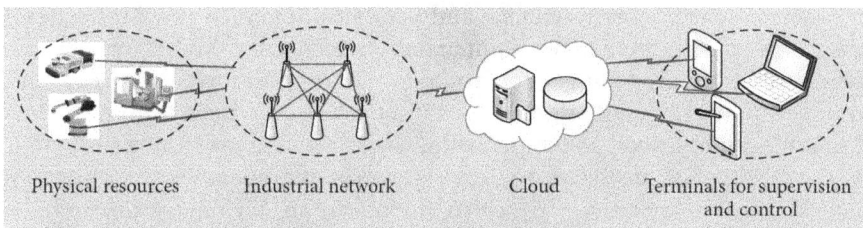

Physical resources   Industrial network   Cloud   Terminals for supervision and control

Figure 7.2 Smart factory framework of Industry 4.0. (Adapted from (S. Wang, Wan, Li, & Zhang, 2016) (CC BY 4.0)).

algorithms may be trained to find flaws and abnormalities in product photos, such as chips, scratches, and fractures. For instance, an ML system that seeks to identify smartphone screen cracks can be trained using a dataset of smartphone picture data (Ding et al., 2020).

Predictive quality control is another way ML is used in quality control. ML algorithms are used by predictive quality control algorithms to assess production data and forecast the possibility of quality problems. For instance, a predictive quality control system may examine data from a manufacturing line and foresee the potential for product flaws, enabling producers to take preventative action (L. Wang, 2019).

7.3.1.1 Example: how machine learning can be applied to improve quality control in a smart factory

Let's say that a smart factory produces electronic components that must meet strict quality standards. The factory uses ML algorithms to analyse data from various sensors that measure the components' quality, such as their dimensions, weight, and electrical properties. The ML algorithm can learn to identify patterns and anomalies in the data that indicate whether a component meets the quality standards or not. For example, it can detect whether a component's dimensions fall outside an acceptable range, whether it has any defects, or whether it meets the electrical specifications. Based on this analysis, the algorithm can flag any components that fail to meet the quality standards, allowing the factory to remove them from the production line before they cause any further problems. This can help reduce waste, improve the overall quality of the products, and avoid any potential issues that could arise from faulty components.

Moreover, by continuously analysing the data, the machine learning algorithm can identify any potential issues with the production process, such as a problem with the raw materials or a flaw in the manufacturing process. This information can help the factory address any problems and improve the production process to prevent similar issues from occurring.

7.3.2 Predictive maintenance

In order to schedule maintenance and prevent unplanned downtime, manufacturers utilize predictive maintenance, which uses ML algorithms to forecast when equipment is likely to break. In order to forecast when maintenance is necessary, ML algorithms can be trained on data from manufacturing machinery, such as temperature, vibration, and pressure data (Çınar et al., 2020). For instance, to forecast when manufacturing equipment, such as a robot arm, is likely to malfunction, an ML algorithm may be trained on data from the machine. To develop predictions about the robot's health, the algorithm can examine performance data, including speed, accuracy, and energy usage. If the programme anticipates the robot failing, the manufacturer may plan maintenance and prevent unplanned downtime (Dalzochio et al., 2020).

7.3.2.1 Example: how predictive maintenance can be applied in a smart factory using ML

Let's say that a smart factory uses ML algorithms to analyse data from sensors installed on a particular machine in the production line. The sensors collect temperature, vibration, and pressure information fed into the ML algorithm. The algorithm uses this data to model the machine's normal behaviour and performance. It then compares real-time sensor data with this model and identifies deviations from normal behaviour.

If the algorithm detects any deviations, it sends an alert to the maintenance team, indicating that the machine will likely fail soon. The maintenance team can then schedule maintenance activities, such as replacing a faulty part or carrying out repairs, before the machine breaks down. By predicting machine failures in advance, the smart factory can avoid unplanned downtime, reduce maintenance costs, and prevent any loss of production. This allows the factory to operate more efficiently and increase its output, resulting in higher profits and customer satisfaction.

7.3.3 Production optimization

Production optimization is making manufacturing more effective to cut costs and boost productivity. ML algorithms can improve production processes by evaluating production data and providing predictions that can guide decision-making. Demand forecasting is one of the applications of ML used most often in production optimization (Ding et al., 2020). Algorithms for demand forecasting employ ML to assess past sales data and anticipate future demand (Wang, 2019).

For instance, an algorithm for demand forecasting may evaluate sales data for a product, like smartphones, and anticipate future demand. Production scheduling is another way that ML is used in production optimization. ML algorithms are used in production scheduling to improve production processes' scheduling while considering aspects like labour availability, equipment availability, and output demand. To ensure that equipment is used effectively and that production demand is satisfied, a production scheduling algorithm can assess production demand and equipment availability and plan production operations (Rai, Tiwari, Ivanov, & Dolgui, 2021).

The use of ML in smart factories has grown in importance since it gives manufacturers a competitive edge and enables them to keep up with the needs of a market that is changing quickly. Smart factories' applications for ML algorithms include quality assurance, proactive maintenance, and production planning. ML algorithms can offer insights that help guide decision-making and enhance the efficiency of production processes, lowering costs and raising productivity by evaluating vast volumes of production data.

7.3.3.1 Example: how ML can be applied to optimize production in a smart factory

Let's say that a smart factory produces a specific product that requires multiple production steps. Each step has several parameters that can affect the quality and quantity of the output, for example, the temperature and humidity in the production area, the speed of the conveyor belt, and the amount of raw material used. The smart factory uses ML algorithms to analyse data from various sensors and machines involved in production. The algorithm can identify patterns and relationships between the production parameters and the quality and quantity of the output. The algorithm then creates a production process model, which it can use to predict the optimal production parameters for each step. The model can be continuously updated as new data is collected, allowing the algorithm to adjust and improve the predictions over time.

Using this information, the smart factory can optimize the production process by adjusting the production parameters in real time to achieve the best possible output quality and quantity. This can help reduce waste, increase efficiency, and improve overall production output. For example, the ML algorithm may determine that a slight increase in the raw material used during a particular production step can significantly improve the output quality. The smart factory can then adjust the raw material to optimize production and enhance the final product's quality.

7.4 APPLICATIONS OF MACHINE LEARNING IN SMART FACTORY

A smart factory's implementation and operation heavily rely on ML. These are a few keys uses of ML in smart factories:

- Maintenance: It may be planned ahead of time, thanks to ML algorithms that can identify when equipment is most likely to break. This can cut down on downtime and lower the chance of production interruptions (Rai et al., 2021).
- Quality control: ML algorithms can examine items and spot flaws in real time, lowering the possibility of sending clients substandard goods.
- Process optimization: Using ML algorithms, industrial processes may be more effective and efficient while minimizing waste. For instance,

ML may decide the best order of operations to build a product using criteria like production speed and material availability.

- Equipment monitoring: ML algorithms can track equipment functioning and identify problems before they become serious. This can lower the likelihood of expensive breakdowns and increase the dependability of the equipment (J. Lee, Azamfar, Singh, & Siahpour, 2020).
- Inventory management: By optimizing inventory levels, ML algorithms may lower the risk of stockouts and surplus inventory (Namir, Labriji, & Ben Lahmar, 2022). For instance, ML can forecast when a specific product is likely out of supply and automatically place an order to refill it (Shao, Liu, Li, Chaudhry, & Yue, 2021).
- Energy optimization: ML algorithms can make the best use of energy in a smart factory, lowering energy costs and the carbon footprint of production. For instance, ML may be used to find the industrial equipment's most energy-efficient operating settings (Jayaprakash, Nagarajan, Prado, Subramanian, & Divakarachari, 2021; Narciso & Martins, 2020).

These are just a few of the many applications of ML in smart factories. By utilizing these technologies, smart factories can improve efficiency, reduce waste, and increase productivity (Nawrocki & Osypanka, 2021; Rai et al., 2021).

7.5 BENEFITS OF IMPLEMENTING MACHINE LEARNING IN SMART FACTORIES

Using ML in smart factories provides several advantages, including greater production, decreased costs, and enhanced efficiency (Rai et al., 2021). The following are some major advantages of ML in smart factories:

7.5.1 Better quality control

Automating quality control procedures using ML algorithms lowers the chance of human mistakes while enhancing productivity. For instance, image analysis algorithms may be trained to find flaws and abnormalities in product photos, like chips, fractures, and scratches, increasing the precision and efficiency of quality control procedures (J. Lee et al., 2020).

7.5.2 Predictive maintenance

Predictive maintenance enables manufacturers to plan maintenance and prevent unplanned downtime by predicting when equipment is likely to break. In order to forecast when maintenance is necessary, predictive maintenance algorithms can examine data from manufacturing machines, such as temperature, vibration, and pressure data. This decreases equipment downtime and lengthens equipment lifespan (Rai et al., 2021).

7.5.3 Production optimization

ML algorithms can improve manufacturing procedures while lowering costs and boosting output. For instance, demand forecasting algorithms may examine past sales data and future project demand, enabling firms to plan their output in an educated manner (Weichert et al., 2019).

7.5.4 Better decision-making

ML algorithms may offer insights into production processes, enabling producers to confidently choose the right production planning, equipment, and quality control strategies. ML algorithms can offer insights that help guide decision-making and boost the effectiveness of manufacturing processes by evaluating vast volumes of production data (Andronie et al., 2021).

7.5.5 Real-time monitoring

ML algorithms can track the status of production lines in real time, giving producers up-to-the-minute information on the efficiency of their operations and the condition of their equipment. Making decisions based on this real-time data can increase the effectiveness of the production process (Rai et al., 2021).

7.6 CHALLENGES OF IMPLEMENTING ML IN SMART FACTORIES

Despite the many advantages of using ML in smart factories, there are still obstacles to be cleared. The following are some of the major issues with ML in smart factories:

7.6.1 Data availability

High-quality data is a major obstacle to implementing ML in smart factories since ML algorithms need much data to be taught. ML algorithms cannot be trained successfully without enough data, which lowers their efficacy (Rai et al., 2021).

7.6.2 Data quality

Another significant issue is the calibre of the data utilized to train ML algorithms. Inaccurate predictions and choices can be made due to bad data, which decreases the efficacy of ML algorithms (Lee et al., 2020).

7.6.3 Integration with current systems

Integrating ML algorithms with current production systems can be difficult and time-consuming. In addition to requiring considerable technical resources and experience, integrating ML algorithms with existing systems can pose new technological hazards, including data security and system incompatibilities (Tambe, Cappelli, & Yakubovich, 2019).

7.6.4 Technological complexity

Implementing ML algorithms can be quite technical, requiring high technical competence. Data analysis, algorithm selection, and model training all demand technical skills, which might be challenging to locate internally (Mittal, Khan, Romero, & Wuest, 2018).

7.6.5 Algorithm selection

Choosing the best ML algorithms for a given application might be challenging since different algorithms are better suited to various data kinds and use. To select the best algorithm, one must have a thorough grasp of ML and the application's particular needs, which might be challenging to acquire internally. In conclusion, incorporating ML into smart factories has several advantages, including higher productivity, cost savings, and enhanced efficiency. Data accessibility, quality, interaction with current systems, technological complexity, and algorithm choice are some of the difficulties that must be solved. Manufacturers must possess the technical know-how, financial means, and dedication to implement ML (Chen et al., 2018; Mittal et al., 2018; Zhong, Xu, Klotz, & Newman, 2017).

7.7 FUTURE OUTLOOK FOR MACHINE LEARNING IN THE INDUSTRIAL SECTOR

The forecast for ML in the industrial sector is favourable, and it is anticipated that the technology will advance and mature while overcoming the obstacles preventing wider implementation. The future of ML in the industrial sector will see several significant trends and advances (Akinosho et al., 2020).

7.7.1 Increased adoption

As ML technology develops and becomes more sophisticated, it is anticipated that more firms will embrace it to boost productivity and competitiveness (Akinosho et al., 2020).

7.7.2 Increased data accessibility

High-quality data is a major obstacle to ML implementation in the industrial sector. This difficulty is anticipated to diminish as data availability increases and data management and analysis technologies advance, making it simpler for businesses to use ML technology.

7.7.3 Better algorithms

It is anticipated that ML algorithms will continue to develop and mature, becoming more sophisticated and able to handle increasingly challenging applications. As algorithms advance, more firms are anticipated to employ technology to boost productivity and competitiveness in the industrial sector (Mosavi et al., 2020).

7.7.4 Integration with existing systems

Integration with current systems is a significant obstacle to ML use in the industrial sector. This difficulty should diminish as integration technology advances, making it simpler for firms to use ML technologies (C. Lee & Lim, 2021).

7.7.5 Data security will get more attention

As ML technology develops, data security and privacy will receive more attention. It will be crucial to ensure that this data is safe and shielded from theft or abuse as more data is gathered and examined. The manufacturing sector has already shown enormous promise for ML, and its potential uses in future smart factories are expected to be considerably more substantial (Zhang & Lu, 2021).

7.7.6 Improved predictive maintenance

As ML algorithms advance, predictive maintenance will become more accurate. ML algorithms, for instance, may examine a far more extensive range of data sources, such as information from sensors and the equipment itself, to forecast when equipment is likely to fail (Akinosho et al., 2020).

7.7.7 Real-time process optimization

ML algorithms will make manufacturing far more flexible and agile by optimizing production processes in real time (Angelopoulos et al., 2020; Uzunidis, Karkazis, & Leligou, 2022). For instance, ML algorithms may automatically modify production schedules in response to changes in demand, lowering the possibility of stockouts or having too much inventory (Kakani, Nguyen, Kumar, Kim, & Pasupuleti, 2020).

7.7.8 Autonomous manufacturing

Autonomous factories with production equipment that can run without human interference will be created using ML algorithms (Antons & Arlinghaus, 2022). This will enable manufacturers to run their facilities around the clock, increasing productivity and lowering labour expenses (C. Wang, Tan, Tor, & Lim, 2020).

7.7.9 Advanced quality control

Improved quality control will be implemented using ML algorithms, including instantaneous product inspection and automated flaw detection. This will lower the possibility of client complaints and assist producers in generating high-quality items (Ryalat, ElMoaqet, & AlFaouri, 2023).

7.7.10 Intelligent supply chain management

The whole supply chain will be optimized using ML algorithms, from sourcing raw materials to distributing completed goods to customers. As a result, producers will save money, operate more effectively, and react swiftly to demand changes (Abbas, Afaq, Ahmed Khan, & Song, 2020).

7.7.11 Virtual and augmented reality

Smart factories will utilize ML algorithms to build virtual and AR environments for training and visualization. For instance, ML algorithms will be used to build digital representations of industrial processes, enabling staff to practise operating machinery securely (Jayaprakash et al., 2021). These are a few prospective directions for ML in smart factories in the Future. As technology develops, ML applications in the industrial sector will undoubtedly be limitless (Li et al., 2021).

The forecast for ML in the industrial sector is favourable, and it is anticipated that the technology will advance and mature while overcoming the obstacles preventing more comprehensive implementation (Zhang & Lu, 2021). More firms are expected to employ ML technology as it offers in the industrial sector to boost productivity and competitiveness. Technology will become increasingly crucial to the industrial sector as it develops, changing manufacturing methods and boosting competitiveness (Andronie et al., 2021; Han et al., 2021).

Table 7.1 lists the key information on the applications of ML in smart factories, its limitations, and future scope. The applications of ML in smart factory shows five vital applications, including predictive maintenance, quality control, production optimization, supply chain optimization, and

Table 7.1 Applications of ML in smart factory, limitations, and future scope.

Applications of Machine Learning in Smart Factory	Limitations	Future Scope
Predictive maintenance	Data quality, high initial costs, complexity, human intervention	Improved data quality, greater accessibility, advancements in algorithms, integration with other technologies
Quality control	Data quality, high initial costs, complexity, human intervention	Improved data quality, greater accessibility, advancements in algorithms, integration with other technologies
Production optimization	Data quality, high initial costs, complexity, human intervention	Improved data quality, greater accessibility, advancements in algorithms, integration with other technologies
Supply chain optimization	Data quality, high initial costs, complexity, human intervention	Improved data quality, greater accessibility, advancements in algorithms, integration with other technologies
Predictive analytics	Data quality, high initial costs, complexity, human intervention	Improved data quality, greater accessibility, advancements in algorithms, integration with other technologies

predictive analytics. The key limitations of ML in smart factories include data quality, high initial costs, complexity, and human intervention. The future scope includes improved data quality, greater accessibility, algorithm advancements, and integration with other technologies. These advancements are expected to enhance the capabilities of ML in smart factories and enable manufacturers to optimize their production processes further.

7.8 IMPLICATIONS OF MACHINE LEARNING IN SMART FACTORIES

Manufacturers can optimize their production processes using ML algorithms to analyse data from various sources, reduce waste, and improve quality. ML algorithms can identify areas where production can be improved, such as identifying bottlenecks in the production process and suggesting ways to optimize them. This can lead to improved efficiency and reduced costs. Predictive maintenance can help reduce downtime by identifying when a machine is likely to fail and scheduling maintenance activities accordingly. This can help manufacturers avoid costly unplanned downtime.

ML algorithms can analyse data from sensors and cameras to detect product defects. This can help manufacturers identify quality issues early, preventing defective products from reaching customers and reducing waste. Manufacturers can make better decisions based on data-driven insights by using ML algorithms for predictive analytics. For example, they can predict demand for products and adjust production accordingly. By leveraging the power of ML in smart factories, manufacturers can increase their competitiveness by reducing costs, improving quality, and delivering products more quickly. Overall, the implications of ML in smart factories are significant, and manufacturers who adopt these technologies will be better positioned to compete in the modern industrial landscape.

7.9 CONCLUDING REMARKS

ML can completely transform the industrial sector by enhancing output, cutting costs, and boosting efficiency. By automating quality control procedures, forecasting equipment failure, streamlining manufacturing procedures, and offering real-time monitoring, ML can significantly influence the industrial sector. Yet, despite its potential advantages, various obstacles must be addressed to deploy ML in the industrial sector. These obstacles include data accessibility, quality, system integration, technological complexity, and algorithm choice.

REFERENCES

Abbas, K., Afaq, M., Ahmed Khan, T., & Song, W.-C. (2020). A blockchain and machine learning-based drug supply chain management and recommendation system for smart pharmaceutical industry. *Electronics, 9*(5). doi:10.3390/electronics9050852

Akinosho, T. D., Oyedele, L. O., Bilal, M., Ajayi, A. O., Delgado, M. D., Akinade, O. O., & Ahmed, A. A. (2020). Deep learning in the construction industry: A review of present status and future innovations. *Journal of Building Engineering, 32*, 101827. doi:10.1016/j.jobe.2020.101827

Alanne, K., & Sierla, S. (2022). An overview of machine learning applications for smart buildings. *Sustainable Cities and Society, 76*, 103445. doi:10.1016/j.scs.2021.103445

Alloghani, M., Al-Jumeily, D., Mustafina, J., Hussain, A., & Aljaaf, A. J. (2020). A systematic review on supervised and unsupervised machine learning algorithms for data science. In M. W. Berry, A. Mohamed, & B. W. Yap (Eds.), *Supervised and Unsupervised Learning for Data Science* (pp. 3–21). Cham: Springer International Publishing.

Andronie, M., Lăzăroiu, G., Iatagan, M., Uță, C., Ştefănescu, R., & Cocoşatu, M. (2021). Artificial intelligence-based decision-making algorithms, internet of things sensing networks, and deep learning-assisted smart process management in cyber-physical production systems. *Electronics, 10*(20). doi:10.3390/electronics10202497

Ang, J. H., Goh, C., Saldivar, A. A., & Li, Y. (2017). Energy-efficient through-life smart design, manufacturing and operation of ships in an Industry 4.0 environment. *Energies, 10*(5). doi:10.3390/en10050610

Angelopoulos, A., Michailidis, E. T., Nomikos, N., Trakadas, P., Hatziefremidis, A., Voliotis, S., & Zahariadis, T. (2020). Tackling faults in the Industry 4.0 Era—a survey of machine-learning solutions and key aspects. *Sensors, 20*(1). doi:10.3390/s20010109

Antons, O., & Arlinghaus, J. C. (2022). Data-driven and autonomous manufacturing control in cyber-physical production systems. *Computers in Industry, 141,* 103711. doi:10.1016/j.compind.2022.103711

Chen, B., Wan, J., Shu, L., Li, P., Mukherjee, M., & Yin, B. (2018). Smart factory of Industry 4.0: Key technologies, application case, and challenges. *IEEE Access, 6,* 6505–6519. doi:10.1109/ACCESS.2017.2783682

Çınar, Z. M., Abdussalam Nuhu, A., Zeeshan, Q., Korhan, O., Asmael, M., & Safaei, B. (2020). Machine learning in predictive maintenance towards sustainable smart manufacturing in Industry 4.0. *Sustainability (Switzerland), 12*(19). doi:10.3390/su12198211

Dalzochio, J., Kunst, R., Pignaton, E., Binotto, A., Sanyal, S., Favilla, J., & Barbosa, J. (2020). Machine learning and reasoning for predictive maintenance in Industry 4.0: Current status and challenges. *Computers in Industry, 123,* 103298. doi:10.1016/j.compind.2020.103298

Ding, H., Gao, R. X., Isaksson, A. J., Landers, R. G., Parisini, T., & Yuan, Y. (2020). State of AI-based monitoring in smart manufacturing and introduction to focused section. *IEEE/ASME Transactions on Mechatronics, 25*(5), 2143–2154. doi:10.1109/TMECH.2020.3022983

Ghobakhloo, M. (2020). Industry 4.0, digitization, and opportunities for sustainability. *Journal of Cleaner Production, 252,* 119869. doi:10.1016/j.jclepro.2019.119869

Han, Y., Liu, S., Cong, D., Geng, Z., Fan, J., Gao, J., & Pan, T. (2021). Resource optimization model using novel extreme learning machine with t-distributed stochastic neighbor embedding: Application to complex industrial processes. *Energy, 225,* 120255. doi:10.1016/j.energy.2021.120255

Jayaprakash, S., Nagarajan, M. D., Prado, R. P., Subramanian, S., & Divakarachari, P. B. (2021). A systematic review of energy management strategies for resource allocation in the cloud: Clustering, optimization and machine learning. *Energies, 14*(17). doi:10.3390/en14175322

Kakani, V., Nguyen, V. H., Kumar, B. P., Kim, H., & Pasupuleti, V. R. (2020). A critical review on computer vision and artificial intelligence in food industry. *Journal of Agriculture and Food Research, 2,* 100033. doi:10.1016/j.jafr.2020.100033

Khan, M. A., Abbas, K., Su'ud, M. M., Salameh, A. A., Alam, M. M., Aman, N., ... Aziz, R. C. (2022). Application of machine learning algorithms for sustainable business management based on macro-economic data: Supervised learning techniques approach. *Sustainability (Switzerland), 14*(16). doi:10.3390/su14169964

Kotsiopoulos, T., Sarigiannidis, P., Ioannidis, D., & Tzovaras, D. (2021). Machine learning and deep learning in smart manufacturing: The smart grid paradigm. *Computer Science Review, 40,* 100341. doi:10.1016/j.cosrev.2020.100341

Lee, C., & Lim, C. (2021). From technological development to social advance: A review of Industry 4.0 through machine learning. *Technological Forecasting and Social Change, 167,* 120653. doi:10.1016/j.techfore.2021.120653

Lee, J., Azamfar, M., Singh, J., & Siahpour, S. (2020). Integration of digital twin and deep learning in cyber-physical systems: Towards smart manufacturing. *IET Collaborative Intelligent Manufacturing, 2*(1), 34–36. doi:10.1049/iet-cim.2020.0009

Li, Y., Lei, G., Bramerdorfer, G., Peng, S., Sun, X., & Zhu, J. (2021). Machine learning for design optimization of electromagnetic devices: Recent developments and future directions. *Applied Sciences, 11*(4). doi:10.3390/app11041627

Malburg, L., Rieder, M.-P., Seiger, R., Klein, P., & Bergmann, R. (2021). Object detection for smart factory processes by machine learning. *Procedia Computer Science, 184,* 581–588. doi:10.1016/j.procs.2021.04.009

Mittal, S., Khan, M. A., Romero, D., & Wuest, T. (2018). A critical review of smart manufacturing & Industry 4.0 maturity models: Implications for small and medium-sized enterprises (SMEs). *Journal of Manufacturing Systems, 49,* 194–214. doi:10.1016/j.jmsy.2018.10.005

Mosavi, A., Faghan, Y., Ghamisi, P., Duan, P., Ardabili, S. F., Salwana, E., & Band, S. S. (2020). Comprehensive review of deep reinforcement learning methods and applications in economics. *Mathematics, 8*(10). doi:10.3390/math8101640

Namir, K., Labriji, H., & Ben Lahmar, E. H. (2022). Decision support tool for dynamic inventory management using machine learning, time series and combinatorial optimization. *Procedia Computer Science, 198,* 423–428. doi:10.1016/j.procs.2021.12.264

Narciso, D. A. C., & Martins, F. G. (2020). Application of machine learning tools for energy efficiency in industry: A review. *Energy Reports, 6,* 1181–1199. doi:10.1016/j.egyr.2020.04.035

Nawrocki, P., & Osypanka, P. (2021). Cloud resource demand prediction using machine learning in the context of QoS parameters. *Journal of Grid Computing, 19*(2), 20. doi:10.1007/s10723-021-09561-3

Oztemel, E., & Gursev, S. (2020). Literature review of Industry 4.0 and related technologies. *Journal of Intelligent Manufacturing, 31*(1), 127–182. doi:10.1007/s10845-018-1433-8

Park, S.-T., Li, G., & Hong, J.-C. (2020). A study on smart factory-based ambient intelligence context-aware intrusion detection system using machine learning. *Journal of Ambient Intelligence and Humanized Computing, 11*(4), 1405–1412. doi:10.1007/s12652-018-0998-6

Pereira, A. C., & Romero, F. (2017). A review of the meanings and the implications of the Industry 4.0 concept. *Procedia Manufacturing, 13,* 1206–1214. doi:10.1016/j.promfg.2017.09.032

Rai, R., Tiwari, M. K., Ivanov, D., & Dolgui, A. (2021). Machine learning in manufacturing and industry 4.0 applications. *International Journal of Production Research, 59*(16), 4773–4778. doi:10.1080/00207543.2021.1956675

Ryalat, M., ElMoaqet, H., & AlFaouri, M. (2023). Design of a smart factory based on cyber-physical systems and Internet of Things towards Industry 4.0. *Applied Sciences, 13*(4). doi:10.3390/app13042156

Sarker, I. H., Kayes, A. S. M., Badsha, S., Alqahtani, H., Watters, P., & Ng, A. (2020). Cybersecurity data science: An overview from machine learning perspective. *Journal of Big Data, 7*(1), 41. doi:10.1186/s40537-020-00318-5

Shao, X.-F., Liu, W., Li, Y., Chaudhry, H. R., & Yue, X.-G. (2021). Multistage implementation framework for smart supply chain management under industry 4.0. *Technological Forecasting and Social Change, 162*, 120354. doi:10.1016/j.techfore.2020.120354

Sony, M., & Naik, S. (2020). Key ingredients for evaluating Industry 4.0 readiness for organizations: A literature review. *Benchmarking: An International Journal, 27*(7), 2213–2232. doi:10.1108/BIJ-09-2018-0284

Tambe, P., Cappelli, P., & Yakubovich, V. (2019). Artificial intelligence in human resources management: Challenges and a path forward. *California Management Review, 61*(4), 15–42. doi:10.1177/0008125619867910

Usuga Cadavid, J. P., Lamouri, S., Grabot, B., Pellerin, R., & Fortin, A. (2020). Machine learning applied in production planning and control: A state-of-the-art in the era of industry 4.0. *Journal of Intelligent Manufacturing, 31*(6), 1531–1558. doi:10.1007/s10845-019-01531-7

Uzunidis, D., Karkazis, P., & Leligou, H. C. (2022, July 20–22). *Machine learning resource optimization enabled by cross layer monitoring. Paper presented at the 2022 13th International Symposium on Communication Systems, Networks and Digital Signal Processing (CSNDSP).*

Wang, C., Tan, X. P., Tor, S. B., & Lim, C. S. (2020). Machine learning in additive manufacturing: State-of-the-art and perspectives. *Additive Manufacturing, 36*, 101538. doi:10.1016/j.addma.2020.101538

Wang, L. (2019). From intelligence science to intelligent manufacturing. *Engineering, 5*(4), 615–618. doi:10.1016/j.eng.2019.04.011

Wang, S., Wan, J., Li, D., & Zhang, C. (2016). Implementing smart factory of Industrie 4.0: An outlook. *International Journal of Distributed Sensor Networks, 12*(1), 3159805. doi:10.1155/2016/3159805

Weichert, D., Link, P., Stoll, A., Rüping, S., Ihlenfeldt, S., & Wrobel, S. (2019). A review of machine learning for the optimization of production processes. *The International Journal of Advanced Manufacturing Technology, 104*(5), 1889–1902. doi:10.1007/s00170-019-03988-5

Zhang, C., & Lu, Y. (2021). Study on artificial intelligence: The state of the art and future prospects. *Journal of Industrial Information Integration, 23*, 100224. doi:10.1016/j.jii.2021.100224

Zhong, R. Y., Xu, X., Klotz, E., & Newman, S. T. (2017). Intelligent manufacturing in the context of Industry 4.0: A review. *Engineering, 3*(5), 616–630. doi:10.1016/J.ENG.2017.05.015

Chapter 8

Role of machine learning in Industry 4.0 applications

A review

Harpreet Kaur Channi

Chandigarh University, Gharuan, Mohali, Punjab India

Raman Kumar

Guru Nanak Dev Engineering College, Ludhiana, India

8.1 INTRODUCTION

Over the past 300 years, humankind has made tremendous industrial production strides. Mechanical advancements using steam and water powered the First Industrial Revolution, while electrification and sophisticated machine tools powered the second, significantly increasing and enhancing output (Azeem et al., 2022; Choudhury, 2021). The Third Industrial Revolution, beginning in the 1950s, paved the way for automated production via widespread usage of semiconductors and, more recently, communication networks (Nagar et al., 2021; Rai et al., 2021). The introduction of artificial intelligence (AI) and ML over the past decade has significantly benefited the manufacturing industry. These benefits include increased quality and productivity, decreased waste and resource consumption, enhanced sustainability, and safer working environments (Jung et al., 2021; Lee & Lim, 2021). AI/ML-based manufacturing may bring various industrial breakthroughs, including issue identification and prediction, effective use of raw materials and resources, heterogeneous extensive data analysis, and networked production facilities (Anastasi et al., 2021; Putnik et al., 2021).

8.1.1 Industry 4.0

The concept of "Industry 4.0," sometimes known as the "Fourth Industrial Revolution," centers on developing manufacturing ecosystems that are more real time, intelligent, interoperable, and autonomous. The advancements in information and communications technology, such as cyber-physical systems (CPS), the Internet of Things (IoT), and cloud computing (CC), which are essential to realizing the vision of Industry 4.0, have piqued the attention of both research institutes and enterprises (Kotsiopoulos et al., 2021; Sarker, 2021). It is the outcome of a CPS that physical and digital systems are combined. Physical components include sensors, control panels, and computers; these components work together and exchange information to provide input to cyber-elements, which are responsible for the management, processing, and decision-making (Teoh et al., 2021). Lastly, IoT enables the reliable and trustworthy real-time connection of many entities, such as sensors, actuators, machines, and robots. The IoT is built on several technologies, including 5G networks, Wi-Fi, M2M deployments, and cloud computing (cloud, fog, edge) (Javaid, Haleem, Singh, Rab, et al., 2021a). However, people should still play a significant role in Industry 4.0 settings by using AI/ML-based decision-making and being equipped with intelligent devices, VR/AR, and a continuous connection to the production process (Bousdekis et al., 2021). Considering these cutting-edge technological advancements, Industry 4.0 has the potential to drastically alter the current industrial production processes to the advantage of industry stakeholders, staff, and customers while also contributing to environmental sustainability (Sajid et al., 2021). Flexibility, competence, real-time self-optimization, and automation are some of the features that may be expected from the many applications envisioned within Industry 4.0 (Rathore et al., 2021). In addition, the most critical pervasive applications are found in manufacturing and new product creation (Bertolini et al., 2021). The fault detection, prediction, and prevention fields play crucial roles in developing these applications. ML algorithms may detect broken or faulty parts or products in real time to reduce downtime. Due to the amount of data, an accurate prediction of the machine's status, RUL, and faults is possible, leading to an appropriate and cost-effective maintenance plan that reduces downtime due to fault occurrence (Mishra & Tyagi, 2022). Humans play a significant part in the manufacturing process of the Industry 4.0 ecosystem, even though working with robots and machines presents considerable difficulty in the workplace (Fan et al., 2021). Algorithms that recognize human actions and help operators make better decisions boost productivity and safety and save ramp-up time (Liu et al., 2021). Capgemini Consulting identifies "smart solutions," "smart innovation," "smart supply chains," and "smart manufacturing" as the four pillars of Industry 4.0 (Bilotta et al., 2021; Elsisi et al., 2021). The same technology underpins both the old and new evolutionary models.

Similarly, McKinsey & Company has divided this set of technologies into four classes: data, computing power, and connection; analytics and intelligence; human–machine interaction; and digital-to-physical conversion. These four categories form the backbone of the digitization framework and naturally cluster together the technologies discussed in the relevant scholarly works (Ahmed et al., 2022). Figure 8.1 summarizes the critical equipment needed for Industry 4.0 to function correctly.

The functions inside current architectures should be mined for needs that technologies can fulfill. These conditions must be met to ensure adaptability, dependability, and interoperability (Rojek et al., 2021). Engineering, planning, production, operations, and logistics will benefit from increased robustness and conformity by Industry 4.0 (Javaid, Haleem, Singh, & Suman, 2022a).

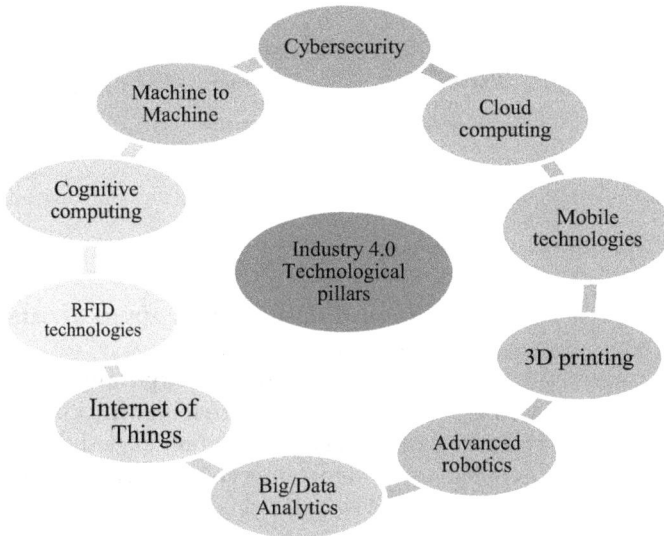

Figure 8.1 Technologies for Industry 4.0.

8.1.2 Machine learning

A considerable amount of data is required for ML algorithm application to industrial operations. Big data analytics (BDA) demonstrates how the emergence of new technology paradigms like CPS and IoT permits the creation of novel data forms (Sader et al., 2022). When gathering information for later analysis, it may be challenging to differentiate between the background noise and the actual data collection (Zheng et al., 2021). Dynamic settings and diverse machine states complicate ML-based fault identification and prevention. Real-time transmission and calculation must be dependable and precise, and security challenges are growing due to greater subsystem interconnectivity (Saturno et al., 2017). Figure 8.2 demonstrates that ML algorithms may be segmented into four categories: supervised, unsupervised, semi-supervised, and reinforcement learning (RL).

- Experts employ supervised learning to train classification and regression algorithms by substituting known outputs for specified inputs. The tagged data is utilized in supervised ML. Standard algorithms include artificial neural networks (ANNs) and support vector machines (SVMs) (Javaid, Haleem, Singh, & Suman, 2021b).
- When no external supervision is present, the algorithm is said to have engaged in unsupervised learning, during which it discovered patterns in previously unseen data (clustering, association rules, self-organized maps). So, unlabeled data are used in the training process of this kind of learning. Principal component analysis (PCA) is the most

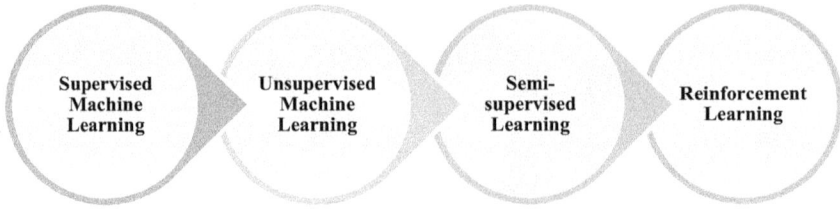

Figure 8.2 Types of machine learning.

well-known and widely applied unsupervised approach and is often utilized in monitoring (Züfle et al., 2022).

- Semi-supervised ML is a kind of ML that incorporates characteristics of both supervised and unsupervised approaches. In order to train its algorithms, it makes use of both labeled and unlabeled datasets. These issues may be avoided using semi-supervised knowledge, which uses both supervised and unsupervised datasets (Chabanet et al., 2022).
- Reinforcement learning is an unsupervised ML operation that checks whether a given action has a positive outcome for a designated performance measure. To solve a problem, RL requires a series of steps to be taken, followed by the results of those steps being tried until the best solution is found. Therefore, RL differs significantly from the categories of learning predicated on using historical data and developing wisdom from prior judgments and incentives (Moosavi, Bakhshi, et al., 2021a).

8.2 LITERATURE REVIEW

ML metamodels were presented to mitigate this expense and use the massive amounts of data now accessible to businesses. Predictive models built into industrial decision support systems must undergo ongoing training to maintain high prediction quality over time (Chabanet et al., 2021). Industry 4.0 has been the subject of a thorough literature study. The thematic analysis identifies the following as essential technologies in the context of Industry 4.0: the Internet of Things (IoT), AI, cloud computing, ML, security, big data, blockchain, deep learning, digitalization, and CPS (Moosavi, Naeni, et al., 2021b).

Since 70–80% of final product costs are established in product development and design processes, note that it is vital to manage and control product costs in the early stages of value creation (Bodendorf & Franke, 2021). Traditional business practice emphasizes the controlling unit of an organization that keeps track of the associated expenditures as they arise. "Industry 4.0" is the next big thing in the manufacturing sector. Massive amounts of information are created due to the Internet's ability to collect data from every device and action through network sensors (Kotsiopoulos et al., 2021). With the introduction of ML and deep learning, two types of AI, manufacturers may analyze the data they collect and gain insight into how to improve their operations (IAI).

One will concentrate on common misconceptions about the future of AI, ML, and the Internet of Things (IoT) (Shirsath et al., 2012). New developments in AI and ML have set the stage for the next generation of manufacturing, called Industry 4.0 (Sambangi et al., 2022). Compared to earlier revolutions observed by humans, Industry 4.0 is moving at a breakneck pace. Industry 4.0 relies heavily on cyber-physical systems and cloud computing. A fresh viewpoint on how AI/DL/ML is used for the whole building lifespan, from the initial idea to the final teardown. In addition, the obstacles in model creation and solutions to solve them are touched upon, and data-gathering tactics that use bright vision and sensors are also highlighted (Baduge et al., 2022).

The approach of transforming black-box models into explainable (and interoperable) classifiers based on semantic principles is proposed through the automatic rebuilding of the training datasets and retraining the decision trees (explainable models) (Terziyan & Vitko, 2022). Another author suggested a sawmill industry-inspired use case to demonstrate these techniques' relevance in fluctuating DT data flows (Chabanet et al., 2021). A semantic platform for heterogeneous IoFT data aggregation that allows interoperability is presented (Cenni, 2022). This dataset included weather and vegetation data collected by Brazilian government sensors and satellite fire data. ML was used to forecast the regions a fire would impact.

Comprehending Industry 4.0 applications and their functions concerning Supply Chain Resilience (SCRes) and Supply Chain Visibility (SCV) was offered. They used partial least square structural equation modeling (PLS-SEM) to examine the potential connections (Qader et al., 2022). Another researcher presented a system disruption monitoring tool for an Industry 4.0 setting. An ML system may produce a localization prediction model when activities are scheduled with the resources' location in mind (Brik et al., 2022). ML and automation have been explored as crucial to the industrial sector's success. ML techniques and high-resolution cameras can automate industrial operations, propelling us toward Industry 4.0. The ML and automation project aims to automate visual inspection using highly accurate fault detection on manufactured parts (Harishyam et al., 2023). New technologies emerging from "Industry 4.0" and those on the horizon of advanced digitalization make it possible to provide more efficient, high-value production and service at a lower cost. A functional healthcare system requires careful oversight of healthcare resources, clinical care procedures, service planning, delivery, and assessment (Karatas et al., 2022).

8.2.1 Applications of machine learning in Industry 4.0

Industry 4.0, sometimes known as the "Fourth Industrial Revolution," describes the widespread use of IT in manufacturing. This is a common way the digital revolution is described inside formal business contexts. The phrase may be broadly interpreted to cover cyber-physical systems, the Internet of Things, cloud computing, data analytics, and ML (Karatas et al., 2022). In this section, we will examine the application of ML to the field of Industry 4.0. With the help of Industry 4.0, intelligent factories may be built (Ammar et al., 2021). Here, cyber-physical technologies are used to keep tabs on the operation. Through the IoT, many systems and human-operated devices and operators can coordinate with one another in real time. Information gathered by cyber-physical systems might be stored centrally in the cloud. ML is used in the last phase of Industry 4.0 to extract previously established patterns from data, gaining prominence as a driving force behind break-throughs in many fields (Bécue et al., 2021) and providing opportunities for businesses to reap benefits such as the following:

- Enhanced quality of products
- Producing with more leeway

The most important applications of ML in the Fourth Industrial Revolution are shown in Figure 8.3.

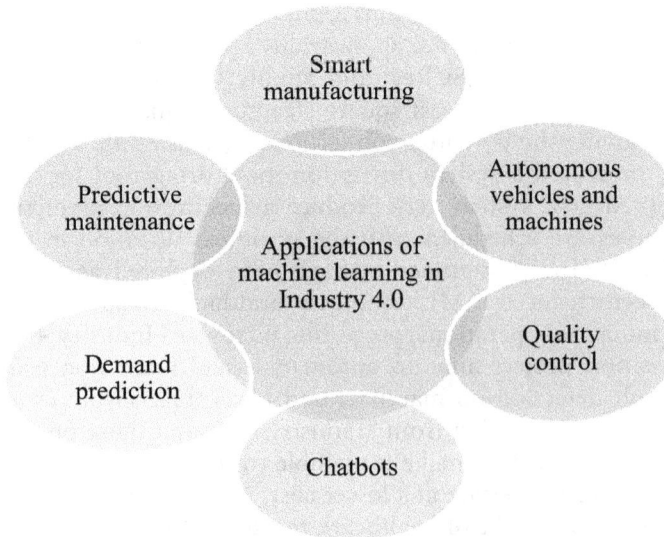

Figure 8.3 Applications of machine learning in Industry 4.0.

8.3 TRANSFORMATION OF THE PRODUCTION PROCESS: "SMART MANUFACTURING"

ML techniques are integrated into manufacturing to comprehend better and optimize production procedures. The data gathered during manufacturing allows this goal to be attained (Jagatheesaperumal et al., 2021; Zobeiry & Humfeld, 2021). The assessment leads to novel procedures with the flexibility to adjust continually to production changes. Since this is the case, separate processes are better monitored and can also be optimized. Intelligent manufacturing automatically puts these improvements into real time (Tsaramirsis et al., 2022).

8.3.1 Autonomous vehicles and machines

One obvious use of AI and ML is in autonomous automobiles and other machinery. In the business world, it frees humans from tedious or dangerous work (Ada et al., 2021).

8.3.2 Quality control

Previously, finished goods inspections were conducted solely as a last step in the manufacturing process. Sensors and ML allow quality to be monitored throughout manufacturing (Moosavi, Bakhshi, et al., 2021a).

8.3.3 Predictive maintenance

Miniaturization and decreases in price have led to a rise in the usage of sensors for machine health monitoring, which has provided users with previously unavailable insights, hence reducing the cost of machine management. So, by installing hundreds or thousands of sensors in the machines, we can monitor their overall health in real time. This data may be used to train ML algorithms to predict when certain system parts will break down (Chabanet et al., 2022).

8.3.4 Demand prediction

Adjusting output to meet demand is a persistent challenge. This is especially true when the outcome is perishable and cannot be kept for later use. The generation of electricity is one such instance. When manufacturing circumstances are reasonable, it is not feasible to put away extra. The emergence of renewable energy sources, whose output is unpredictable, also adds complexity to administration. Using ML, we can better satisfy the ideal energy demand in this market. The anticipated demand may be calculated by looking at past energy use trends. However, it is possible to estimate renewable energy generation using weather data (Meindl et al., 2021).

8.3.5 Chatbots

Chatbots allow consumers to discuss with a computer system through text or voice. They lower the threshold for gaining access to data and labor-saving tools. Their primary function is to assist clients before, during, and after a transaction or to screen inquiries before forwarding them to more specialized human advisers. They help you save money and provide support around the clock, but you can also use them to study the most prevalent issues (Malik et al., 2022).

8.4 TOP MACHINE LEARNING STARTUPS IMPACTING INDUSTRY 4.0

8.4.1 Rejig Digital provides predictive analytics

Indian firm Rejig Digital creates tailored industrial ML solutions to advance the Fourth Industrial Revolution, founded in 2020 with headquarters in Ahmedabad, India. The company's data analytics solution, which employs big data and ML algorithms, can evaluate the enormous volumes of data generated by enterprise resource planning (ERP) software, industrial sensors, and connected devices. Predictive analytics in real time is one of the many benefits of the data analytics solution, which can deal with issues like budget overruns, broken equipment, and inefficient production methods. Electronics manufacturers, power companies, utility providers, and banks may all save money using the startup's solution (Varshney et al., 2021).

8.4.2 Siali Tech advances process optimization

Siali Tech, headquartered in Spain, employs computer vision and deep learning to enhance operations anywhere they occur. It all started in Santander, Spain, in 2018. The company's deep learning platform, Inspector, can adapt to any business environment and provide solutions to improve efficiency by learning visual tasks. The Inspector improves packing and shipping processes, aids in inventory management by finding irregular items, monitors equipment and employees, and uncovers other opportunities for improvement. The startup's platform may be tailored to serve industries as diverse as the food and beverage, automobile, construction, consumer electronics, and logistics industries (Srivastava et al., 2022).

8.4.3 Nexocraft enables predictive maintenance

The German company Nexocraft developed an online platform called Graphicx.io that uses ML to evaluate the condition of industrial machines and make maintenance plans. It has been operational since 2016 and is situated in Bonn, Germany. The Visualization and Evaluation tool from

Graphicx.io can monitor industrial equipment and provide Industrial Health Scoring since it learns the optimal operating conditions for all systems. It compiles information from all of the attached sensors and uses an automated analysis of the data to determine whether or not there have been any significant shifts in performance. Nexocraft's solutions allow businesses to pick ML models tailored to key performance metrics for their unique industrial processes while minimizing equipment downtime (Oluyisola et al., 2022).

8.4.4 mSense facilitates fault detection

Established in the United States, it is a firm that works on a real-time verification platform throughout manufacturing processes to provide consistent quality control and innovation. It opened in the US city of Milpitas in 2018 after its founding. To diagnose and repair mechanical failures, the company's ASDL platform uses ML algorithms for acoustics, vibration, and vision and sensors linked to the IIoT. The technology allows rapid problem diagnosis locally, alleviates pressure on cloud storage, and may be implemented with little to no additional development time. The ASDL platform enhances the capabilities of businesses such as healthcare, automobile, utilities, and insurance by providing industry-specific AcousticDL, TactileDL, and VisionDL modules (Narayanamurthy & Tortorella, 2021).

8.4.5 Tignis accelerates process simulations

PAICe Maker is a powerful AI and ML software solution developed by Tignis, a startup company in the United States. The Seattle-based firm's headquarters opened in 2017. By manipulating a wide range of control parameters, it can simulate a wide range of production situations, allowing for the design and optimization of industrial components and processes. AI algorithms run on the platform's edge nodes to regulate manufacturing processes and shorten feedback loops. PAICe Maker is software that helps manufacturers and process engineers enhance their physical assets and procedures over time. It also speeds up the simulation process and improves the plant's quality and output using actual data (Ferreira et al., 2022).

8.5 INDUSTRY 4.0'S IMPACT

The worldwide spread of COVID-19 has impacted every area of the economy. The manufacturing process is always the same. However, digital transformation and applied ML in the industry may help mitigate some adverse effects. Enterprise monitoring systems are only one example of an Industry 4.0 solution that allows for the hands-free monitoring of several factories by a central team. This information is then saved in the cloud, which can be accessed and analyzed through dashboards (Paturi & Cheruku, 2021). The advantages are apparent. It is safer for workers and more convenient for businesses when the health of a plant can be monitored appropriately without the need for a packed floor, especially when experts advise social separation.

Several older industries and companies have been sluggish to adopt Industry 4.0 due to their distrust of new technology. The unique challenges of 2020 will strain many, but the advantages of an Industry 4.0 setting in a globally distributed workforce are becoming more evident. While most businesses are likely attempting to dig down and save costs, don't be surprised if enterprises in historically resistive areas start thriving by using ML and other Industry 4.0 technology, such as remote monitoring systems (Sadiq et al., 2021). The future relevance of ML in the industrial environment cannot be emphasized, especially given the enormous growth potential of Industry 4.0 and the numerous success stories from intelligent factories already permeating the market. These effects are also not minor or inconsequential (Sahal et al., 2021). Savings and efficiency benefits in the hundreds of millions of dollars may be realized immediately by adopting the methods and tools of Industry 4.0. Manufacturing is an area to keep an eye on for rapid expansion due to Industry 4.0's continuous development and acceptance.

8.5.1 Impact of ML on Industry 4.0 as compared to previous prediction approaches

Despite being around for a while, ML algorithms have recently seen a surge in interest because of AI's rise. In particular, modern deep learning models fuel the most cutting-edge AI programs. ML platforms are a highly competitive area of enterprise technology (Ghobakhloo et al., 2021). Offering comprehensive platform services for all stages of the ML lifecycle, including data collection, data preparation, data classification, model building, training, and application deployment, is a strategy that is being utilized by a large number of corporations, including Amazon, Google, Microsoft, IBM, and others, to attract customers and compete for market share. As the manufacturing sector undergoes a digital transition, the Fourth Industrial Revolution, or Industry 4.0, is the next logical step. Because some see it as the Fourth Industrial Revolution, the number "4.0" has been attached. Using cutting-edge information and communications technology in industrial production is at the heart of this concept (Diez-Olivan et al., 2019). It is common practice to release a new version of a product if substantial upgrades have been made. These developments, collectively known as "Industry 4.0," set the Fourth Industrial Revolution apart from its predecessors. In Figure 8.4, we can see the many processes throughout manufacturing. Integrating intelligent technology into supply chains and enterprises is a hallmark of Industry 4.0, the fourth major revamp of contemporary industry (Bajic et al., 2018). Its revolutionary impact on the modern sector has earned it the moniker "the Fourth Industrial Revolution." When applied to goods, processes, and

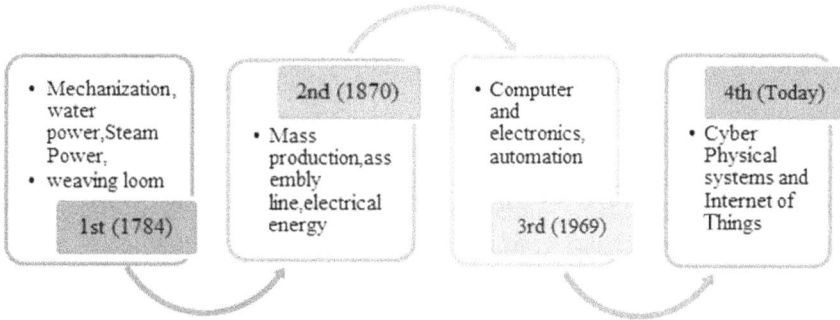

Figure 8.4 Different stages of industrial production.

people, the interconnected technology in the Industry 4.0 movement satisfies the requirement for connectivity and real-time information (Ustundag & Cevikcan, 2018). Rather than relying on centralized control systems like those employed in the early 20th century, the factories of Industry 4.0 make choices on their own using decentralized innovative technology.

Over many hundred years, the Industrial Revolution reshaped the world as we know it. With each new period comes the development of cutting-edge technologies and the accompanying reorganization of economic sectors to suit them (Paolanti et al., 2018). Before Industry 4.0, there were three significant shifts in the industrial sector:

- First Industrial Revolution: Modern industry was born when this emphasis shifted from human labor to instruments and steam engines. Throughout the 18th century, economies throughout the globe started to acknowledge and appreciate the importance of technology in production.
- Second Industrial Revolution: Industry 2.0 saw a dramatic increase in productivity and output due to the widespread adoption of new technologies and the widespread availability of previously inaccessible materials, such as steel and energy, in production.
- Third Industrial Revolution: Industry 3.0, which occurred in the 1950s and is often considered the beginning of automation, served as a model for the later creation of supply chains, assembly lines, and computerization of production procedures.
- The modern ramifications of Industry 4.0 surpass those of its forerunners. With the support of the Internet of Things (IoT), access to real-time data, intelligent computing, and the introduction of cyber-physical systems, it elevates the focus on digital technology from previous decades to a new level.

(Candanedo et al., 2018)

Three-dimensional printing, nanotechnology, quantum computing, cloud computing, and energy optimization are further examples of instrumental technologies. Industry 4.0 has enabled unparalleled accuracy and productivity in supply chains and production by integrating the finest of current technologies into a corporate environment (Ciolacu et al., 2017). This has resulted in several advantages on many different markets' supply and demand sides. The advancements in technology that underpin Industry 4.0 are crucial to the sector's continued success. We are now closer than ever before to a point when the physical, digital, and biological lines are invisible because of the enormous variety of technological solutions organizations use today (Miškuf & Zolotová, 2016).

Industry 4.0 is also known as Industrial IoT for its reliance on wireless cyber-physical systems. Every component, from automated sensors to mechanical gear, is linked to a distributed processing unit for data. Greater supply chain visibility, access to information, and management are the most significant advantages of such extensive interconnectedness (Preuveneers & Ilie-Zudor, 2017). Manufacturers can foresee potential issues and conduct routine maintenance to avoid them. Providers may assess market conditions and adjust product delivery accordingly. Likewise, stores may monitor their possessions and fine-tune logistical factors like quality and quantity (Candanedo et al., 2018).

8.6 FIVE INDUSTRIES THAT SIGNIFICANTLY DEPEND ON AI AND ML

The development of mind-reading technology is moving the globe closer to discovering the quickest, most efficient method to do anything. ML and AI are on the rise, and businesses across sectors are using these tools to conduct more in-depth market research and streamline daily processes (Frank et al., 2019). Technologies like this employ algorithms built to generate high-level processing for a computer, allowing it to learn and accomplish the work without explicit programming. Companies such as Coca-Cola and Heineken, American Express and Experian, Alexa and Chat-Bot, Babylon and Freenome, and Deep Blue and DeepMind are all in this category. Recently, DeepMind defeated the finest Go player in the world, following in the footsteps of the groundbreaking Deep Blue (Javaid, Haleem, Singh, Suman, et al., 2022b).

8.6.1 Transportation

Transportation technology is cutting-edge, and it has no limits. Many businesses in the automation industry are taking notice of the growing popularity of using AI and ML algorithms for traffic prediction, monitoring, and management. The primary focus is on autonomous vehicles. Although they are still experimental, self-driving vehicles represent the future transportation system (Mishra & Tyagi, 2022). Algorithms are being created for such complicated goods so that a design may take advantage of new features, such as evaluating and optimizing data acquired from various sources and adapting routing, mapping, and navigation strategies to their circumstances. The ability to analyze multiple objects' surroundings and process the path of control makes computer vision and sensor fusion essential applications in autonomous vehicles (Rao et al., 2022). As a result, the system can make informed judgments and adapt its actions in real time based on data and information gleaned from various sources. It's no surprise that carmakers like BMW and Volkswagen and logistics firms like those in Japan employ teams of AI specialists to work on such diverse projects.

8.6.2 Healthcare

The healthcare sector is being catapulted forward by AI technology, which can analyze and comprehend complex life science data at a far higher level than ever. AI-driven diagnostics analyze patient data to accurately detect illness and provide potential treatments. While AI has the potential to discover medical issues much more quickly than a human sense, it does not intend to elevate new ways of treatment (Javaid, Haleem, Singh, & Suman, 2022a). Kohonen's Neural Network, a popular neural network, is a simple example of a self-organizing network that collects and displays data more nuancedly. Researchers in biological sciences may benefit from the speed with which AI and ML enable them to read, which aids in their ability to learn about the root causes of illness and to create more accurate diagnostic tools (Oluyisola et al., 2022).

Leading healthcare technology business Philips embraces adaptive intelligence, a platform combining AI and other technologies to better serve the industry's demands. Many business companies, such as Babylon Health, Infravision, Freenome, and others, concentrate on developing analytical models to forecast and identify illnesses before beginning therapy (Jabbar et al., 2022).

8.6.3 Finance

The banking and investment sector have been early adopters of AI. With such a large quantity of high-quality data available, AI algorithms are increasingly used throughout the industry, from data collection and management to transaction processing and investment analysis. AI provides strategic moves for company growth. AI-driven financial services and solutions are the keys to a safe and successful company (Park et al., 2022). Its primary purpose is to protect financial institutions from fraudulent activity while providing the services their customers need. The financial industry places a high value on these innovations, used to develop chatbots and conversational interfaces that provide customers with the precise data they need. With more and more people doing business online, the need for cybersecurity has become paramount. Financial institutions such as JP Morgan, American Express, Experian, and many more have worked with well-known firms and professionals to raise industry standards and mitigate potential risks to their clients, investors, and institutions (Ahmad et al., 2022). As one of several cybersecurity firms, Palo Alto Network has released Magnifier, which uses ML algorithms to uncover and eliminate threats hiding in plain sight inside an organization.

8.6.4 Agriculture

With the adaptability of AI and ML, the widespread use of automation in agriculture might usher in a new era of prosperity. Since there is a lack of human labor, these technologies are employed primarily to create autonomous robots for tasks like weed detection, soil monitoring, and disease diagnosis in agriculture. Farmers may boost output with robotics (Khan & Javaid, 2022). As a bonus, these robots may aid with weed detection and identification, allowing the user to take the necessary action. A plant's health may also be assessed since these diagnostic tools might identify problems with the soil or harvest problems. Google is currently developing a tool that can detect plant illness, identify the specific disease that has impacted a plant, determine the causes of the sickness, and provide treatments for the problem. ICRISAT and Microsoft developed the Sowing App to increase crop yields for farmers (Haleem et al., 2022). This information is helpful for farmers since it tells them when to start preparing the land, when to plant seeds, how long to soak them, how much fertilizer to use, how to deal with pests and diseases, and how much water to use. The influence increased yields by 30% per hectare (Pillai & Srivastava, 2022).

8.6.5 Retail and customer service

Every business sector looks to the market to learn whether a potential product can make it in the real world before investing time and money into producing it. AI and ML allow businesses to better understand their customers' wants and needs by analyzing data and making informed predictions. Future e-commerce and brick-and-mortar stores will rely heavily on AI and ML technology (Xie, 2022). AI and ML are employed in retail since it is difficult to comprehend what consumers are looking for and locate relevant items; as a result, they help customers find what they're looking for more quickly and easily. However, it helps create new platforms, such as 3D clothing design and virtual try-on. Conversational chatbots and digital support show how technology alters the service industry. The end users are assisted, and an AI-driven customer service representative answers questions. Thanks to developments in ML and related technologies like natural language processing, everyday conversations may be conducted in real time, complete with quick, insightful replies based on analyzing the request and identifying the purpose and entities involved (Han & Trimi, 2022). To provide accurate details about his shows, music, and location, DJ Hardwell is the first EDM artist to use a chatbot.

8.7 THE FUTURE OF INDUSTRY 4.0

A lack of robust manufacturing would prevent developing nations from fully embracing Industry 4.0. They need to diversify their production base into industries that need higher technical sophistication. Fostering a culture of participation, promoting promising industries, strengthening innovation infrastructure, fostering consistency between science, technology, and information (STI) policies and other social and economic initiatives, and fostering consistency between STI policies and other social and economic initiatives are all essential responsibilities of the state (Ahmed et al., 2022). Governments should encourage affordable, high-quality Internet access and help businesses, particularly small- and medium-sized businesses (SMEs), develop digital literacy. In addition, they should foster the circumstances necessary for the rollout of Industry 4.0 in production. There are many examples of this, including developing national strategies to guide the coordinated deployment of Industry 4.0, establishing a multistakeholder mechanism to institutionalize a participatory approach to foster Industry 4.0, and the growth of international cooperation to accelerate the transfer of technology and know-how. Governments should increase private sector understanding of Industry 4.0, encourage investments, and ease financing for the deployment of Industry 4.0 to promote its acceptance. Policymakers should focus on trade and global value chain shifts and how they can influence developing world labor forces (Mishra & Tyagi, 2022). Workers who cannot be retrained or rehired and thereby lose employment should depend on more robust social protection systems.

The rate of today's Industry 4.0 innovations is unprecedented and that much is certain. Due to the near-exponential increase in computer and processing capacity from one year to the next, we should expect even greater revolutions in an already revolutionary era. Quantum computing is one technology that has enormous potential in the not-too-distant future. Unlike traditional computers, Quantum computers execute computing using quantum phenomena like superposition. Quantum computers, in a nutshell, are liberated from the electrical constraints of the past (Rao et al., 2022). As we go ahead, the convergence of technological advancements like quantum computing and cutting-edge technical systems will undoubtedly propel Industry 4.0 to new heights. The potential benefits of Industry 4.0 are enormous, ranging from assisting contemporary industry in breaking free of traditional limits to allowing for the effective growth of firms and supply networks (Javaid, Haleem, Singh, & Suman, 2022a).

There's much hope for the future of ML. Right now, ML applications are the driving force behind almost every industry. The healthcare sector, search engines, digital marketing, and the education sector are just a few examples of industries that stand to gain significantly. With the potential for computers to take over human labor, ML's value to a business or organization might be questioned. Machine learning emerges as the biggest boon of AI for efficiently achieving goals (He et al., 2023). Unpredictable progress is being made in computer vision and natural language processing (NLP). Both are present in modern technology, from face recognition apps on smartphones to automatic language translators and driverless automobiles. What was once the stuff of science fiction is now becoming a reality. The pervasiveness of ML now makes it difficult to picture a world without it (Ukoba et al., 2023). Our forecast for ML's future growth in 2022 and beyond is as follows.

Industry 4.0 is making its way into companies and factories worldwide through enhancing ML. AI has spawned a subject known as ML, which enables computers and computer programs to teach themselves new abilities and improve their performance. The field of manufacturing has seen substantial progress in optimization as a direct consequence of using ML (Naqvi et al., 2023). A "smart factory" may be created if the production process is optimized to use ML. Smart factories, the digitalized equivalent of conventional factories, rely heavily on automation and data collection to streamline production. Businesses can make more informed decisions, thanks to the superior analytics made possible by this data aggregation. One of the most apparent advantages of ML in manufacturing is its potential to increase productivity without necessitating significant changes to the current infrastructure. One such instance is mistake detection in real time (Li et al., 2023). Through the use of intelligent devices on the factory floor, smart factories can quickly evaluate product quality. Video streaming devices with ML capabilities can monitor production from when an idea is conceived to the final shipment. After that, the machine may examine each frame rapidly for defects. With the use of ML, video analysis (VA), and real-time performance monitoring (LTM), engineers may get accurate and timely insights to address issues (Bhagwan & Evans, 2023).

8.8 CHALLENGES OF IMPLEMENTING INDUSTRY 4.0

The term "Industry 4.0" describes the most recent trend in the automation of factories and the exchange of information. Cloud computing, cognitive computing, the Internet of Things, and cyber-physical systems are some examples of these emerging technologies. Because of Industry 4.0, there will be establishments referred to as "smart factories." Inside the intelligent factory are cyber-physical systems that monitor the physical processes, replicate the natural environment, and make decisions independently (Chen et al., 2023). With the proliferation of IoT devices, merging the digital and physical realms will speed up. Germany was the birthplace of the Industry 4.0 idea, which has now spread throughout the globe. The purpose of Industry 4.0 is to develop a production system that is more flexible and responsive to market demands (Soori et al., 2023).

While the benefits of Industry 4.0 are clear, adopting it is difficult. A recent poll found that simulation and computation, automation, and data management are the top three areas where businesses need improvement. A lack of software tools, automation in design and production, data transfer and collecting, and production data management are all problems shared by these disciplines. For Industry 4.0 to realize its full potential, several obstacles must be overcome (Ventayen, 2023).

8.9 CONCLUSIONS

Industry 4.0 is gaining traction in organizations and manufacturing facilities, thanks primarily to ML. ML, a subfield of AI, is a technique for teaching computers to become more innovative. Applying ML to Industry 4.0 is a growing trend that will have far-reaching effects. The transition will impact multinational enterprises, mom-and-pop shops, and other local establishments. As data gathering and processing gear become more affordable, more people will have access to this technology. The companies included in this chapter that are examples of Industry 4.0 emphasize condition monitoring, quality inspection, and preventative analytics. While these technologies are essential for the market's growth, they are simply the tip of the technological innovation iceberg.

REFERENCES

Ada, N., Kazancoglu, Y., Sezer, M. D., Ede-Senturk, C., Ozer, I., & Ram, M. (2021). Analyzing barriers of circular food supply chains and proposing industry 4.0 solutions. *Sustainability*, *13*(12), 6812.

Ahmad, T., Zhu, H., Zhang, D., Tariq, R., Bassam, A., Ullah, F., AlGhamdi, A. S., & Alshamrani, S. S. (2022). Energetics systems and artificial intelligence: Applications of industry 4.0. *Energy Reports*, *8*, 334–361.

Ahmed, I., Jeon, G., & Piccialli, F. (2022). From artificial intelligence to explainable artificial intelligence in industry 4.0: A survey on what, how, and where. *IEEE Transactions on Industrial Informatics*, *18*(8), 5031–5042.

Ammar, M., Haleem, A., Javaid, M., Walia, R., & Bahl, S. (2021). Improving material quality management and manufacturing organizations system through Industry 4.0 technologies. *Materials Today: Proceedings*, *45*, 5089–5096.

Anastasi, S., Madonna, M., & Monica, L. (2021). Implications of embedded artificial intelligence-machine learning on safety of machinery. *Procedia Computer Science, 180*, 338–343.

Azeem, M., Haleem, A., & Javaid, M. (2022). Symbiotic relationship between machine learning and Industry 4.0: A review. *Journal of Industrial Integration and Management, 7*(03), 401–433.

Baduge, S. K., Thilakarathna, S., Perera, J. S., Arashpour, M., Sharafi, P., Teodosio, B., Shringi, A., & Mendis, P. (2022). Artificial intelligence and smart vision for building and construction 4.0: Machine and deep learning methods and applications. *Automation in Construction, 141*, 104440.

Bajic, B., Cosic, I., Lazarevic, M., Sremcev, N., & Rikalovic, A. (2018). Machine learning techniques for smart manufacturing: Applications and challenges in industry 4.0. *Department of Industrial Engineering and Management Novi Sad, Serbia, 29, 29*, 29–38.

Bécue, A., Praça, I., & Gama, J. (2021). Artificial intelligence, cyber-threats and Industry 4.0: Challenges and opportunities. *Artificial Intelligence Review, 54*(5), 3849–3886.

Bertolini, M., Mezzogori, D., Neroni, M., & Zammori, F. (2021). Machine Learning for industrial applications: A comprehensive literature review. *Expert Systems with Applications, 175*, 114820.

Bhagwan, N., & Evans, M. (2023). A review of industry 4.0 technologies used in the production of energy in China, Germany, and South Africa. *Renewable and Sustainable Energy Reviews, 173*, 113075.

Bilotta, E., Bertacchini, F., Gabriele, L., Giglio, S., Pantano, P. S., & Romita, T. (2021). Industry 4.0 technologies in tourism education: Nurturing students to think with technology. *Journal of Hospitality, Leisure, Sport & Tourism Education, 29*, 100275.

Bodendorf, F., & Franke, J. (2021). A machine learning approach to estimate product costs in the early product design phase: A use case from the automotive industry. *Procedia CIRP, 100*, 643–648.

Bousdekis, A., Lepenioti, K., Apostolou, D., & Mentzas, G. (2021). A review of data-driven decision-making methods for industry 4.0 maintenance applications. *Electronics, 10*(7), 828.

Brik, B., Boutiba, K., & Ksentini, A. (2022). Deep learning for B5G open radio access network: Evolution, survey, case studies, and challenges. *IEEE Open Journal of the Communications Society, 3*, 228–250.

Candanedo, I. S., Nieves, E. H., González, S. R., Martín, M. T. S., & Briones, A. G. (2018). *Machine learning predictive model for industry 4.0. Knowledge Management in Organizations: 13th international conference, KMO 2018*, Žilina, Slovakia, August 6–10, 2018, Proceedings 13,

Cenni, C. (2022). *On the Proximate Links Between Object Play and Tool Use in the Context of Stone Handling Behavior in Balinese Long-tailed Macaques Lethbridge*. University of Lethbridge, [Dept. of Psychology].

Chabanet, S., Bril El-Haouzi, H., Morin, M., Gaudreault, J., & Thomas, P. (2022). Toward digital twins for sawmill production planning and control: benefits, opportunities, and challenges. *International Journal of Production Research, 61*(7), 2190–2213. https://doi.org/10.1080/00207543.2022.2068086

Chabanet, S., El-Haouzi, H. B., & Thomas, P. (2021). Coupling digital simulation and machine learning metamodel through an active learning approach in Industry 4.0 context. *Computers in Industry, 133*, 103529.

Chen, M.-W., Chang, M.-S., Mao, Y., Hu, S., & Kung, C.-C. (2023). Machine learning in the evaluation and prediction models of biochar application: A review. *Science Progress*, *106*(1), 00368504221148842.

Choudhury, A. (2021). The role of machine learning algorithms in materials science: A state of art review on industry 4.0. *Archives of Computational Methods in Engineering*, *28*(5), 3361–3381.

Ciolacu, M., Tehrani, A. F., Beer, R., & Popp, H. (2017). Education 4.0—Fostering student's performance with machine learning methods. *2017 IEEE 23rd international symposium for design and technology in electronic packaging (SIITME)*,

Diez-Olivan, A., Del Ser, J., Galar, D., & Sierra, B. (2019). Data fusion and machine learning for industrial prognosis: Trends and perspectives towards Industry 4.0. *Information Fusion*, *50*, 92–111.

Elsisi, M., Tran, M.-Q., Mahmoud, K., Mansour, D.-E. A., Lehtonen, M., & Darwish, M. M. (2021). Towards secured online monitoring for digitalized GIS against cyber-attacks based on IoT and machine learning. *IEEE Access*, *9*, 78415–78427.

Fan, W., Chen, Y., Li, J., Sun, Y., Feng, J., Hassanin, H., & Sareh, P. (2021). Machine learning applied to the design and inspection of reinforced concrete bridges: Resilient methods and emerging applications. *Structures*, *33*, 3954–3963. https://doi.org/10.1016/j.istruc.2021.06.110

Ferreira, L., Putnik, G. D., Varela, M. L. R., Manupati, V. K., Lopes, N., Cunha, M., Alves, C., & Castro, H. (2022). A framework for collaborative practices platforms for humans and machines in Industry 4.0–oriented smart and sustainable manufacturing environments. In *Smart and Sustainable Manufacturing Systems for Industry 4.0* (pp. 1–24). CRC Press.

Frank, A. G., Dalenogare, L. S., & Ayala, N. F. (2019). Industry 4.0 technologies: Implementation patterns in manufacturing companies. *International Journal of Production Economics*, *210*, 15–26.

Ghobakhloo, M., Iranmanesh, M., Grybauskas, A., Vilkas, M., & Petraitė, M. (2021). Industry 4.0, innovation, and sustainable development: A systematic review and a roadmap to sustainable innovation. *Business Strategy and the Environment*, *30*(8), 4237–4257.

Haleem, A., Javaid, M., Singh, R. P., Rab, S., & Suman, R. (2022). Perspectives of cybersecurity for ameliorative Industry 4.0 era: a review-based framework. *Industrial Robot*, *49*(3), 582–597. https://doi.org/10.1108/IR-10-2021-0243

Han, H., & Trimi, S. (2022). Towards a data science platform for improving SME collaboration through Industry 4.0 technologies. *Technological Forecasting and Social Change*, *174*, 121242.

Harishyam, B., Jenarthanan, M., Rishivanth, R., Rajesh, R., & Girish, N. S. (2023). Visual inspection of mechanical components using visual imaging and machine learning. *Materials Today: Proceedings*, *72*, 2557–2563.

He, C., Zhang, C., Bian, T., Jiao, K., Su, W., Wu, K.-J., & Su, A. (2023). A review on artificial intelligence enabled design, synthesis, and process optimization of chemical products for Industry 4.0. *Processes*, *11*(2), 330.

Jabbar, A., Abbasi, Q. H., Anjum, N., Kalsoom, T., Ramzan, N., Ahmed, S., Rafi-ul-Shan, P. M., Falade, O. P., Imran, M. A., & Ur Rehman, M. (2022). Millimeter-wave smart antenna solutions for URLLC in Industry 4.0 and beyond. *Sensors*, *22*(7), 2688.

Jagatheesaperumal, S. K., Rahouti, M., Ahmad, K., Al-Fuqaha, A., & Guizani, M. (2021). The duo of artificial intelligence and big data for industry 4.0: Applications, techniques, challenges, and future research directions. *IEEE Internet of Things Journal, 9*(15), 12861–12885.

Javaid, M., Haleem, A., Singh, R. P., Rab, S., & Suman, R. (2021a). Significance of sensors for industry 4.0: Roles, capabilities, and applications. *Sensors International, 2*, 100110.

Javaid, M., Haleem, A., Singh, R. P., & Suman, R. (2021b). Significant applications of big data in Industry 4.0. *Journal of Industrial Integration and Management, 6*(04), 429–447.

Javaid, M., Haleem, A., Singh, R. P., & Suman, R. (2022a). Artificial intelligence applications for industry 4.0: A literature-based study. *Journal of Industrial Integration and Management, 7*(01), 83–111.

Javaid, M., Haleem, A., Singh, R. P., Suman, R., & Gonzalez, E. S. (2022b). Understanding the adoption of Industry 4.0 technologies in improving environmental sustainability. *Sustainable Operations and Computers, 3*, 203–217. https://doi.org/10.1016/j.susoc.2022.01.008

Jung, H., Jeon, J., Choi, D., & Park, J.-Y. (2021). Application of machine learning techniques in injection molding quality prediction: Implications on sustainable manufacturing industry. *Sustainability, 13*(8), 4120.

Karatas, M., Eriskin, L., Deveci, M., Pamucar, D., & Garg, H. (2022). Big data for healthcare Industry 4.0: Applications, challenges and future perspectives. *Expert Systems with Applications, 200*, 116912. https://doi.org/10.1016/j.eswa.2022.116912.

Khan, I. H., & Javaid, M. (2022). Role of Internet of Things (IoT) in adoption of Industry 4.0. *Journal of Industrial Integration and Management, 7*(04), 515–533.

Kotsiopoulos, T., Sarigiannidis, P., Ioannidis, D., & Tzovaras, D. (2021). Machine learning and deep learning in smart manufacturing: The smart grid paradigm. *Computer Science Review, 40*, 100341.

Lee, C., & Lim, C. (2021). From technological development to social advance: A review of Industry 4.0 through machine learning. *Technological Forecasting and Social Change, 167*, 120653.

Li, C., Zheng, P., Yin, Y., Wang, B., & Wang, L. (2023). Deep reinforcement learning in smart manufacturing: A review and prospects. *CIRP Journal of Manufacturing Science and Technology, 40*, 75–101.

Liu, C., Zheng, P., & Xu, X. (2021). Digitalisation and servitisation of machine tools in the era of Industry 4.0: a review. *International Journal of Production Research, 61*(12), 4069–4101. https://doi.org/10.1080/00207543.2021.1969462

Malik, N., Tripathi, S. N., Kar, A. K., & Gupta, S. (2022). Impact of artificial intelligence on employees working in Industry 4.0 led organizations. *International Journal of Manpower, 43*(2), 334–354.

Meindl, B., Ayala, N. F., Mendonça, J., & Frank, A. G. (2021). The four smarts of Industry 4.0: Evolution of ten years of research and future perspectives. *Technological Forecasting and Social Change, 168*, 120784.

Mishra, S., & Tyagi, A. K. (2022). The role of machine learning techniques in Internet of things-based cloud applications. *Artificial Intelligence-based Internet of Things Systems*, 105–135.

Miškuf, M., & Zolotová, I. (2016). *Comparison between multi-class classifiers and deep learning with focus on industry 4.0. 2016 cybernetics & informatics (K&I).*

Moosavi, J., Bakhshi, J., & Martek, I. (2021a). The application of industry 4.0 technologies in pandemic management: Literature review and case study. *Healthcare Analytics*, 1, 100008.

Moosavi, J., Naeni, L. M., Fathollahi-Fard, A. M., & Fiore, U. (2021b). Blockchain in supply chain management: A review, bibliometric, and network analysis. *Environmental Science and Pollution Research*, 1–15. https://doi.org/10.1007/s11356-021-13094-3

Nagar, D., Raghav, S., Bhardwaj, A., Kumar, R., Singh, P. L., & Sindhwani, R. (2021). Machine learning: Best way to sustain the supply chain in the era of industry 4.0. *Materials Today: Proceedings*, 47, 3676–3682.

Naqvi, S. R., Ullah, Z., Taqvi, S. A. A., Khan, M. N. A., Farooq, W., Mehran, M. T., Juchelková, D., & Štěpanec, L. (2023). Applications of machine learning in thermochemical conversion of biomass-A review. *Fuel*, 332, 126055.

Narayanamurthy, G., & Tortorella, G. (2021). Impact of COVID-19 outbreak on employee performance–moderating role of industry 4.0 base technologies. *International Journal of Production Economics*, 234, 108075.

Oluyisola, O. E., Bhalla, S., Sgarbossa, F., & Strandhagen, J. O. (2022). Designing and developing smart production planning and control systems in the industry 4.0 era: a methodology and case study. *Journal of Intelligent Manufacturing*, 33(1), 311–332.

Paolanti, M., Romeo, L., Felicetti, A., Mancini, A., Frontoni, E., & Loncarski, J. (2018). *Machine learning approach for predictive maintenance in industry 4.0. 2018 14th IEEE/ASME international conference on mechatronic and embedded systems and applications (MESA)*.

Park, K. T., Jeon, S.-W., & Noh, S. D. (2022). Digital twin application with horizontal coordination for reinforcement-learning-based production control in a re-entrant job shop. *International Journal of Production Research*, 60(7), 2151–2167.

Paturi, U. M. R., & Cheruku, S. (2021). Application and performance of machine learning techniques in manufacturing sector from the past two decades: A review. *Materials Today: Proceedings*, 38, 2392–2401.

Pillai, R., & Srivastava, K. B. (2022). Smart HRM 4.0 for achieving organizational performance: a dynamic capability view perspective. *International Journal of Productivity and Performance Management*(ahead-of-print). https://doi.org/10.1108/IJPPM-04-2022-0174

Preuveneers, D., & Ilie-Zudor, E. (2017). The intelligent industry of the future: A survey on emerging trends, research challenges and opportunities in Industry 4.0. *Journal of Ambient Intelligence and Smart Environments*, 9(3), 287–298.

Putnik, G. D., Shah, V., Putnik, Z., & Ferreira, L. (2021). Machine learning in cyber-physical systems and manufacturing singularity–it does not mean total automation, human is still in the centre: Part II–I n-CPS and a view from community on Industry 4.0 impact on society. *Journal of Machine Engineering*, 21, 133–153.

Qader, G., Junaid, M., Abbas, Q., & Mubarik, M. S. (2022). Industry 4.0 enables supply chain resilience and supply chain performance. *Technological Forecasting and Social Change*, 185, 122026.

Rai, R., Tiwari, M. K., Ivanov, D., & Dolgui, A. (2021). Machine learning in manufacturing and industry 4.0 applications. *International Journal of Production Research*, 59, 4773–4778. https://doi.org/10.1080/00207543.2021.1956675

Rao, T. V. N., Gaddam, A., Kurni, M., & Saritha, K. (2022). Reliance on artificial intelligence, machine learning and deep learning in the Era of industry 4.0. *Smart Healthcare System Design: Security and Privacy Aspects*, 281–299.

Rathore, M. M., Shah, S. A., Shukla, D., Bentafat, E., & Bakiras, S. (2021). The role of ai, machine learning, and big data in digital twinning: A systematic literature review, challenges, and opportunities. *IEEE Access*, *9*, 32030–32052.

Rojek, I., Macko, M., Mikołajewski, D., Sága, M., & Burczyński, T. (2021). Modern methods in the field of machine modelling and simulation as a research and practical issue related to Industry 4.0. *Bulletin of the Polish Academy of Sciences. Technical Sciences*, *69*(2).

Sader, S., Husti, I., & Daroczi, M. (2022). A review of quality 4.0: Definitions, features, technologies, applications, and challenges. *Total Quality Management & Business Excellence*, *33*(9–10), 1164–1182.

Sadiq, A., Anwar, M., Butt, R. A., Masud, F., Shahzad, M. K., Naseem, S., & Younas, M. (2021). A review of phishing attacks and countermeasures for Internet of things-based smart business applications in industry 4.0. *Human Behavior and Emerging Technologies*, *3*(5), 854–864.

Sahal, R., Alsamhi, S. H., Breslin, J. G., Brown, K. N., & Ali, M. I. (2021). Digital twins collaboration for automatic erratic operational data detection in industry 4.0. *Applied Sciences*, *11*(7), 3186.

Sajid, S., Haleem, A., Bahl, S., Javaid, M., Goyal, T., & Mittal, M. (2021). Data science applications for predictive maintenance and materials science in context to Industry 4.0. *Materials Today: Proceedings*, *45*, 4898–4905.

Sambangi, S., Gondi, L., & Aljawarneh, S. (2022). A feature similarity machine learning model for DDoS attack detection in modern network environments for industry 4.0. *Computers and Electrical Engineering*, *100*, 107955.

Sarker, I. H. (2021). Machine learning: Algorithms, real-world applications and research directions. *SN Computer Science*, *2*(3), 160.

Saturno, M., Pertel, V. M., Deschamps, F., & Loures, E. (2017). *Proposal of an automation solutions architecture for Industry 4.0. 24th international conference on production research*.

Shirsath, S. E., Patange, S., Kadam, R., Mane, M., & Jadhav, K. (2012). Structure refinement, cation site location, spectral and elastic properties of Zn2+ substituted NiFe2O4. *Journal of Molecular Structure*, *1024*, 77–83.

Soori, M., Arezoo, B., & Dastres, R. (2023). Machine learning and artificial intelligence in CNC machine tools, a review. *Sustainable Manufacturing and Service Economics*, 100009. https://doi.org/10.1016/j.smse.2023.100009

Srivastava, Y., Ganguli, S., Suman Rajest, S., & Regin, R. (2022). Smart HR competencies and their applications in Industry 4.0. *A Fusion of Artificial Intelligence and Internet of Things for Emerging Cyber Systems*, 293–315.

Teoh, Y. K., Gill, S. S., & Parlikad, A. K. (2021). IoT and fog computing based predictive maintenance model for effective asset management in industry 4.0 using machine learning. *IEEE Internet of Things Journal*.

Terziyan, V., & Vitko, O. (2022). Explainable AI for Industry 4.0: Semantic representation of deep learning models. *Procedia Computer Science*, *200*, 216–226.

Tsaramirsis, G., Kantaros, A., Al-Darraji, I., Piromalis, D., Apostolopoulos, C., Pavlopoulou, A., Alrammal, M., Ismail, Z., Buhari, S. M., & Stojmenovic, M. (2022). A modern approach towards an industry 4.0 model: From driving technologies to management. *Journal of Sensors*, *2022*, https://doi.org/10.1155/2022/5023011

Ukoba, K., Kunene, T. J., Harmse, P., Lukong, V. T., & Chien Jen, T. (2023). The role of renewable energy sources and Industry 4.0 focus for Africa: A review. *Applied Sciences*, *13*(2), 1074.

Ustundag, A., & Cevikcan, E. (2018). *Industry 4.0: Managing the Digital Transformation*. Springer.

Varshney, A., Garg, N., Nagla, K., Nair, T., Jaiswal, S., Yadav, S., & Aswal, D. (2021). Challenges in sensors technology for industry 4.0 for futuristic metrological applications. *Mapan*, *36*(2), 215–226.

Ventayen, R. J. M. (2023). Industry 4.0: Discovering trends through a bibliometric analysis from 2011 up to 2021. *Available at SSRN 4324832*.

Xie, Q. (2022). Machine learning in human resource system of intelligent manufacturing industry. *Enterprise Information Systems*, *16*(2), 264–284.

Zheng, T., Ardolino, M., Bacchetti, A., & Perona, M. (2021). The applications of Industry 4.0 technologies in manufacturing context: A systematic literature review. *International journal of production research*, *59*(6), 1922–1954.

Zobeiry, N., & Humfeld, K. D. (2021). A physics-informed machine learning approach for solving heat transfer equation in advanced manufacturing and engineering applications. *Engineering Applications of Artificial Intelligence*, *101*, 104232.

Züfle, M., Moog, F., Lesch, V., Krupitzer, C., & Kounev, S. (2022). A machine learning-based workflow for automatic detection of anomalies in machine tools. *ISA Transactions*, *125*, 445–458.

Chapter 9

Supervised learning-assisted models for the manufacturing of sustainable composites

Aditi Mahajanan, Inderdeep Singh, and Navneet Arora
Indian Institute of Technology Roorkee, Roorkee, India

9.1 INTRODUCTION

In materials technology, polymer composites have witnessed exponential growth in terms of applications in the past decade. Compared to conventional materials (metals and non-metals), composites have a high strength-to-weight ratio and adaptability, allowing design flexibility resulting in energy-efficient and sustainable products. Generally, polymer matrix composites (PMCs) are defined as the combination of two or more physically and chemically distinct materials that have enhanced properties compared to base materials. A few highlighting properties of PMCs are high strength, low density, high corrosion resistance, easy processing, dimensional stability, durability, and lower thermal conductivity. These versatile, tailorable, and multifunctional properties have helped PMCs to capture the materials market in various industries, including aerospace, marine, automobile, microelectronics, civil, healthcare, and others. Due to the rising concern and awareness for environmentally sustainable materials, researchers have shifted their research interest toward sustainable composites that are eco-friendly and biodegradable. In contrast to conventional composite materials, biocomposite materials promote sustainable ecosystems since they are synthesized using natural fibers, resins, and oils. Researchers are widely investigating natural fiber-reinforced polymer composites (NFRPCs) for their potential applications in the aerospace and automobile sector as an environment-friendly alternative with additional benefits of low weight, cost, and better crash resistance. Few studies show the carbon storage potential of NFRPCs, mostly wood, hemp, and bamboo fiber composites (*Natural Fibres Show Outstandingly Low CO$_2$ Footprint Compared to Glass and Mineral Fibres – Renewable Carbon News*, n.d.; Pervaiz & Sain, 2003; Ramesh et al., 2017; Sun et al., 2020). This shows that the NFRPCs have a modest contribution to reducing carbon footprints.

However, a few limitations of NFRPCs are low water resistivity, high moisture sensitivity, flammability, and low melting temperature. Sometimes, selected natural fiber is incompatible with the polymer. In that case, fiber is pre-treated chemically or the polymer is co-blended with a compatibilizer to improve the interfacial strength – the poor interfacial bonding results in

DOI: 10.1201/9781003453567-9

poor mechanical properties of the specimen. High complexity is involved during composites processing due to the various factors influencing the composites' characteristics. One of the significant factors is the natural fiber properties, which are not consistent for the same type of fiber but are dependent on geographical and climate conditions. Other factors include the manufacturing process, resin properties, fiber properties (shape, aspect ratio, orientation), chemical composition, and additives (Gairola et al., 2022; Kaushik et al., 2022; Mahajan et al., 2022).

Global companies are now adopting manufacturing practices focusing on the materials and production processes utilizing sustainable resources while minimizing the environmental footprints. This involves an appropriate selection of material, composition, processing, and their interactions to achieve the desired attributes that result in a complex structure during the product development process. For a designer, referring to the emerging literature on sustainable materials or performing trial-and-error experiments wastes time and resources. The increasing demand and system complexity of composites require an exhaustive investigation of their behavior exposed under different conditions. Most modern organizations seek materials and processes that are adaptable and affordable while meeting several conflicting objectives. It's crucial to avoid limiting the polymer composites to a specific target by focusing only on a small number of process variables while considering the majority as constant to investigate these materials' broader utility. Instead, all the process variables must be deemed concurrently for optimization, which would significantly add to the material design area. This calls for computational intelligence for optimization and highly accurate predictive tools built on the data analyses from the extracted information. The applications of machine learning (ML) in materials sciences have proved to be the most reliable and sustainable technique beyond traditional computational methods' capabilities (Choudhury, 2021).

ML, a branch of artificial intelligence, uses statistical and probabilistic techniques to determine viable solutions by learning from the results of experiments. It determines the hidden patterns by assessing the fed data, referred to as model training. The predictions for the new data are made based on these patterns. Through their various studies, researchers have proved ML to be a promising tool for novel materials modeling to achieve tunable multifunctional properties. The authors (Rath et al., 2022) have discussed the importance of ML while designing the product and its manufacturing. Manufacturing companies are adopting sustainable practices that align with the *Sustainable Development Goal 12: Sustainable Consumption and Production* to endure the competitive market and address the ecological challenges. ML facilitates the manufacturing of cost-effective products through automation, optimal material and process selection, predictive maintenance, and reduced production cycle time while maintaining the high quality of the deliverables. This has helped in the cost and time reduction that would be spent on the experimentation. ML has overcome the limitations of traditional statistical approaches for decision-making by finding the underlying nonlinear complex relationships between the various manufacturing process variables (Almanei et al., 2021). A study outlined the importance of machine learning algorithms at each manufacturing stage of PMCs. From the selection of material and process, sample characterization, and process parameter optimization to the end user application, it finds usage at many stages to manufacture an effective product (Pattnaik et al., 2021). The problem type and the dataset available determine the ML algorithm to be utilized. The data required for training in ML for investigating emerging materials can be curated computationally or experimentally. Incorporating data analytics and machine learning techniques in manufacturing has led to Industry 4.0.

9.2 BASIC FRAMEWORK OF MACHINE LEARNING

ML uses the data and generates a program to perform a task. ML algorithms are broadly classified into supervised, unsupervised, and reinforcement learning. Supervised learning has gained more importance in the field of PMCs for the classification or prediction of various properties. In this, well-labeled data is used by the algorithms based on which the predictions are made. Unsupervised learning is used for unlabeled data for data visualization, dimensionality reduction, and outlier detection, which uses clustering or association rules (Mueller et al., 2016). The basic structure of supervised learning is shown in Figure 9.1.

9.2.1 Data acquisition

Firstly, the problem definition should be clear for generating the model. Based on the problem, relevant explanatory variables are considered for data acquisition from the available resources. Data refers to the raw facts and figures in structured or unstructured form. The acquisition indicates acquiring data for the given task at hand. As any ML model requires data for training, the first step is data collection from relevant sources.

9.2.2 Data cleaning and pre-processing

The retrieved raw data contains missing, inconsistent, and erroneous data; outliers, mixed-type data (numerical and categorical), making it inappropriate to be fed to the model. Even the best ML algorithms do not perform as intended without high-quality data and data cleansing. Therefore, cleaning and data preparation needs much time to be invested. The pre-processing step helps convert the crude data to clean feasible data by eliminating data irregularities and normalizing them (Figure 9.2). The data is pre-processed

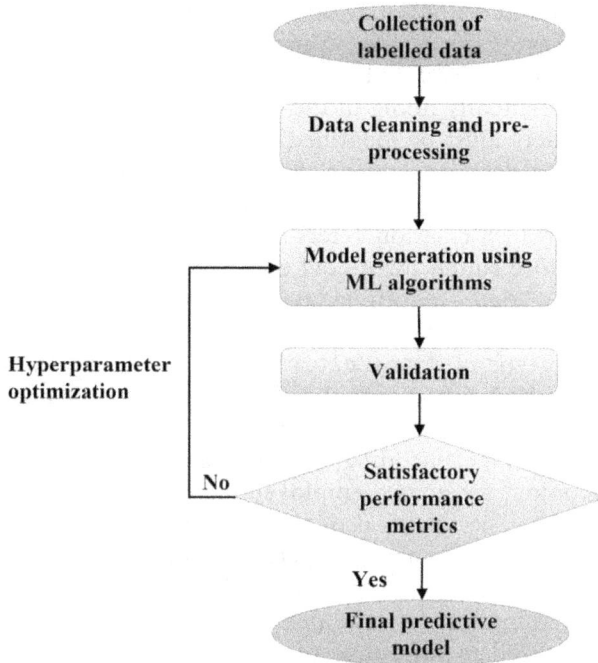

Figure 9.1 Basic framework of machine learning.

Data Cleaning
- Handling missing data, noisy data,
- Removal of outliers,
- Minimizing duplication

Data Integration
Integrating multi sourced data

Data Reduction
- Removing redundant data,
- Feature extraction

Data Transformation
- Data smoothing,
- Aggregation,
- Discretization,
- Normalization,
- Data type conversion

Figure 9.2 Major steps in data cleaning and pre-processing.

so that several ML algorithms may be tested for prediction, and the best one is selected based on performance evaluation metrics.

Missing data can be dealt with in one of three ways: by disregarding the missing record, manually filling in the values, or filling in with calculated values. The outliers must be identified and treated by removing or filling them. During the data integration, data from various sources are converted to the standard form/units and made consistent with assimilating the data. Data transformation refers to transforming unstructured raw data into one more suited for model construction and data discovery in general. The term "data smoothing" refers to methods for removing undesirable noise or behaviors from data. Sometimes, the features/variables contain a mixed type of data (i.e., categorical, text, and numerical). But most ML algorithms are incapable of handling such datatype. Hence, datatype conversion is necessary for interpretation by the machine. Various techniques, such as label encoding, one hot transformation, and backward difference encoding, are mainly employed for categorical data conversion to numerical type. The selection of these techniques depends on various factors and determines the model's performance. The data is normalized to compare the features with different scales of measurement. This step improves the training stability and performance of the model. The data is generally normalized in the range of 0 to 1. More input features often make a predictive modeling task more challenging, more generally referred to as the curse of dimensionality. Essential non-redundant variables with minimum correlation are selected to reduce the number of input features. The other way for feature reduction is feature extraction, which builds valuable variables by constructing the combination of existing features. The process of abstraction of significant attributes for achieving higher accuracy is generally termed feature engineering (Iguyon & Elisseeff, 2003; Mueller et al., 2016).

9.3 MODEL GENERATION

Following the data preparation and pre-processing, a suitable learning algorithm is selected. The appropriate selection depends on the features' data type and the problem objective, that is, regression or classification problem. The algorithm learns the complex relationship between the output and the input variables. Majorly used algorithms for regression problem includes linear regression (LR), ensemble methods, neural networks (NN), gaussian process regression (GPR), and support vector regression (SVR). The classification problems are solved by NN, logistic regression (LoR), support vector machine (SVM), Naïve Bayes, k-nearest neighbor, decision tree (DT), and random forest. After an appropriate selection of the learning algorithm, the processed data is cataloged into three subgroups: training set, validation set, and testing set. Suitable subset creation is necessary for building an efficient model. The chosen subset must be precisely reflective of the complete analytic domain. Using a training dataset, the model acquires the ability to process the input. The validation set aids in fine-tuning algorithmic parameters, enhancing performance by avoiding the overfitting issue. At times, validation is considered part of the training phase. The test set is assured to contain input data matched with outputs that have been manually confirmed to be accurate. This ideal set is employed to test outcomes and assess the success of the resulting model based on the performance metrics (Reitermanová, 2010).

During the training process, choosing the optimum hyperparameters allows the learning algorithm to learn the optimal parameters that accurately map the input features to the response variable. Hyperparameters control the learning process and the model parameters that result from it. It includes data subset creation, learning rate for optimization algorithm, number of hidden layers, activation function, and epochs in NN. They may vary for different algorithms and remain the same unless tweaked to enhance model performance. Parameters are an internal part of the model, including coefficients in LR and LoR, weights, and biases for the NN. Fine-tuning the hyperparameters to improve the model's performance is known as hyperparameter optimization (Probst et al., 2019).

9.3.1 Performance metric evaluation

Once the model is developed, it is essential to evaluate its accuracy for unseen data to select the best-performing model. Confusion matrix, the area under curve, accuracy, recall, and precision are a few metrics for evaluating classification models. For assessing regression model performance, mean absolute error (MAE), mean squared error (MSE), root mean square error (RMSE), and coefficient of determination (R^2) are used. Models are also evaluated based on the training speed and prediction time (Liao et al., 2012).

Various ML algorithms and model evaluation metrics are discussed in detail in later sections of the chapter.

9.4 MACHINE LEARNING APPLICATIONS IN PMCS

Researchers are constantly probing to utilize the potential of ML best to advance the field of PMCs through optimizing material properties for specific applications and developing optimum designs. The research literature on such activities for the most exploited supervised learning techniques is discussed here. When the labeled feature/dependent variable is numerical, the problem is said to be regression. In the classification task, the dependent variable is categorical. The material scientists have successfully used various ML models to predict PMC behavior under varying situations. Table 9.1 summarizes similar research studies.

Table 9.1 Research studies based on the utilization of ML algorithms in the domain of sustainable composites.

Material	Database Source	ML Algorithm	Input	Outputs	References
CaCO$_3$-impregnated coir polyester composites	Experimental	ANN	Fiber length and diameter, CaCO$_3$ filler content	Tensile, flexural, and impact strength	(Jayabal et al., 2013)
Natural fiber	Literature	Stepwise regression	Density, TS, TM, Elongation at break	Material selection	(Muhammad Noryani et al., 2019)
Nano TiO$_2$-coated cotton composites	Experimental	MLR, MLP	Amount of chemicals, reaction time	Functional properties	(Amor et al., 2021)
NFRPCs	Experimental	LR	Particle size, weight fraction	Tensile, flexural, and impact strength	(Prabu et al., 2018)
Bamboo-wood composites	Experimental	MLR, ANN	Specimen characteristic parameters, spectrum parameters	Elastic modulus	(You et al., 2022)
Sisal fiber reinforced paper pulp composites	Experimental	MLR	Fiber volume fraction, fiber length, composite thickness	Sound absorption properties	(Tholkappiyana et al., 2015)
Wood PLA composite	Experimental	MLR	Layer height, infill density, filling pattern	Tensile properties, energy absorption, toughness	(Vigneshwaran & Venkateshwaran, 2019)
NFRPCs	Literature	MLR	Tensile and flexural properties	Material selection	(M. Noryani et al., 2018)
Wood fiber composite	Experimental	CNN	2D patches of 400*400 pixels	Fiber bundle segmentation	(Kibleur et al., 2022)
Starch, wood-reinforced polyvinyl alcohol	Experimental	MLP	Plasticizer content, cross-linking agent, blowing agent	Tensile strength	(Zeng et al., 2019)

Material	Type	ML models	Input parameters	Output properties	Reference
Palm, luffa fiber-reinforced PMCs	Experimental	ANN, MLR, ANFIS, SVR	Fiber, matrix, fiber volume fraction	Tensile strength, strain, elastic modulus	(Alhijazi et al., 2022)
Rice straw mineral composite slabs	Experimental	DT, fully connected cascade ANN	Predicted water absorption, slab density, board thickness, fiber length, mixing ratio	Water absorption, swelling thickness	(Madhappan et al., 2021)
Bio-nanocomposites	Experimental	DT, AdaBoost	Constituent compositions	Fracture toughness	(Daghigh et al., 2020)
Bamboo-wood composites	Experimental	SVM, ANN	Characteristic parameters of image processing	Short span shear stress, flexural strength, Young's modulus	(Jiang et al., 2021)
Jute fiber-reinforced concrete composites	Computational	SVR, ANN	Water cement ratio, length, and volume of jute fiber	Compressive and tensile strength	(Sultana et al., 2020)
NFRPCs	Literature	Multi-SVM	SEM images	Damage classification	(Rajiv et al., 2022)
Cornstalk fibers tied with clay	Experimental	SVR	Frequency, water fraction, thickness	Sound absorption coefficient	(Ciaburro et al., 2021)
Virgin high-density polyethylene	Experimental	LR, LoR, DT, SVM, RF, ANN, AdaBoost	Film component proportions, process parameters	Tensile strength	(Altarazi et al., 2019)
Miscanthus lightweight concrete	Experimental	GPR	Specimen constituents, curing time, and fiber pre-treatment condition	Compressive strength	(Pereira Dias et al., 2021)
Roselle/sisal fiber hybrid polyester composite	Experimental	ANN, Nonlinear regression	Feed rate, cutting speed, drill diameter	Thrust force, torque	(Athijayamani et al., 2010)
Coir and Sisal fiber-reinforced polyester composites	Experimental	ANN	Specimen thickness, fiber volume fraction	Tensile, flexural, impact strength	(Keerthi Gowda et al., 2020)

(Continued)

Table 9.1 (Continued)

Material	Database Source	ML Algorithm	Input	Outputs	References
Bamboo-wood composite container flooring	Experimental	ANN	Layup configuration, density, direction, thickness	Rupture modulus, elastic modulus	(Su et al., 2022)
Marble dust/kenaf hybrid polyester composites	Experimental	ANN	Sliding distance, velocity; load; filler content	Tensile properties, flexural and impact strength, sliding wear rate	(Nayak & Satapathy, 2021)
Cotton fiber PP composites	Experimental	Deep NN	Initiation energy, propagation energy, total energy, ductility index, tensile properties, net weight of fiber	Fiber weight fraction	(Kazi et al., 2020a)
Cotton fiber PVC composites	Experimental	Deep NN	Fiber content, normalized specific energy, load	Displacement	(Kazi et al., 2020b)

9.4.1 Linear regression (LR)

The simplest and most widely used ML algorithm, LR, is used for predictive analysis to predict continuous or numeric variables. Simple LR defines the linear relationship between a dependent and an independent variable. When there is more than one independent variable, it is said to be multiple linear regression (MLR). The line that best fits the data points describing the dependent variable variation with the independent variables represents the line of regression, which can be computationally written as given in Equation 9.1.

$$y = a_0 + a_1x_1 + a_2x_2 + \varepsilon \tag{9.1}$$

where y is the dependent/target variable, x_1, x_2 are independent/predictor variables, a_1, a_2 are the LR coefficients, a_0 is an intercept, and ε is the error term, also known as the residual. The optimal value of coefficients is generally determined by the gradient descent method, which minimizes the MSE. The error terms are assumed to follow the normal distribution (Chen & Gu, 2019). The simplicity and higher explicability nature of LR help find its broader applications. Prabu et al. (2018) investigated the effect of red mud particle size and its weight percentage reinforced in the NFRPCs on the tensile, flexural, and impact strength using the MLR. A study by Vigneshwaran and Venkateshwaran (2019) deployed MLR to predict the mechanical properties of wood-reinforced PLA composites fabricated by fused deposition modeling.

9.4.2 Logistic regression (LoR)

LoR, based on the sigmoid function, is a statistical technique that can be applied to regression tasks and binary classification. LoR is generally applied for predicting discrete values (i.e., classification problems). The response variable lies between 0 and 1, which can be determined using Equation 9.2. An S-shaped logistic curve is fitted instead of a straight line, which returns the probabilistic values for the discrete class labels (Chen & Gu, 2019).

$$log\left(\frac{y}{1-y}\right) = a_0 + a_1x_1 + a_2x_2 + \varepsilon \tag{9.2}$$

Cao et al. (2004) assessed the thermal stability of different polymers performed by thermogravimetric analysis using the LoR. In a study, Osburg et al. (2016) classified customer interest in purchasing wood-based polymer composites using LoR.

9.4.3 Decision tree

DT algorithm is built recursively by breaking down the dataset into smaller chunks resulting in a tree structure form with decision nodes and leaf nodes. The topmost node is the root node. The decision node represents the decision rules that branch further into the decision node, and the leaf node (terminal node) is the outcome of the task. The division into smaller subsets is grounded on their homogeneity. The algorithm can perform both the regression and classification tasks dealing with categorical and numerical data. The optimal tree size is necessary as the giant tree increases overfitting while the smaller tree may be poorly learned. Tree pruning helps determine the optimal size without compromising the model's accuracy (Quinlan, 1986). In a recent study, Gupta et al. (2022) analyzed the effect of functional parameters (grit size, applied load, sliding distance, and fiber weight fraction) on the wear behavior of sisal-reinforced epoxy composite using the DT algorithm.

9.4.4 Support vector machine

SVM, used for classification and regression tasks, creates the best decision boundary, known as a hyperplane, that segregates the multidimensional space into categories in a way that results in a minimum error. The hyperplane is created by maximizing the tolerance margin, meaning the maximum distance between the support vectors. For binary classification, a linear SVM classifier is used. The nonlinear dataset is converted into higher dimensional space using the nonlinear SVM kernel functions (e.g., gaussian, sigmoid, and polynomial) that help in the linear separability (Sharma et al., 2022). A study (Sultana et al., 2020) used the SVR approach to predict jute fiber-reinforced concrete composites' compressive and tensile strength based on the three parameters. Rajiv et al. (2022) classified the damage forms in the NFRPCs by employing the SVM approach based on the SEM images collected from the literature.

9.4.5 Gaussian process regression

GPR is a commonly used probabilistic non-parametric Bayesian learning technique for regression and classification problems that usually works on smaller datasets. In contrast to other standard supervised machine learning techniques, which learn exact values for each function parameter, the Bayesian method estimates a probability distribution for all possible values. Kernel functions make GPR a very effective tool for modeling nonlinear data. GPR presumes that independent and dependent variables have similar Gaussian distribution. The GPR expression can be expressed as given in Equation 9.3. The functions $\mu(x)$ and $k(x, x')$ represent the mean and covariance, respectively (Liao et al., 2012).

$$y = GP\big(\mu\big(x\big), k\big(x, x'\big)\big) \tag{9.3}$$

Pereira Dias et al. (2021) predicted the compressive strength of miscanthus lightweight concrete through the GPR approach based on the inputs of sample constituents, curing time, and fiber pre-treatment condition. The GPR model used two kernel functions: squared exponential and rational quadratic. The rational quadratic kernel-based GPR outperformed, resulting in minimum prediction errors. In a recent study, Okuyama et al. (2022) employed the radial basis kernel-based GPR model to predict the bending modulus of composite materials.

9.4.6 Ensemble methods

The ensemble method is the amalgamation of different models that extracts the benefits of all the base models, thus resulting in an optimal predictive model. The three dominating methods in this domain are bagging, stacking, and boosting. Bagging (Bootstrap AGGregatING) is generally used for aggregating DT predictions fitted on different training data samples. Random forest is a widely used bagging ensemble method. The creation of dissimilar datasets from the training data is known as bootstrapping, and the process of averaging the predictions from different trained models is referred to as aggregating. The stacking method uses the meta-learner to combine the multiple weak learners' predictions based on the same training dataset. Boosting method enables each model learner to learn from the prior member's wrong predictions in a sequence of weak learners, ensuring better performance than the individual learners (Dietterich, 2000). In a study, Wang et al. (2020) applied the random forest approach to characterize the cutting processes for NFRPCs with different fiber microstructures based on the time-frequency features of acoustic emission signals. A study (Daghigh et al., 2020) employed an adaptive boosting regressor to predict the fracture toughness of multiscale bio-nanocomposites based on the different combinations of constituents (relative proportions).

9.4.7 Neural networks

As we have seen, LR is limited to developing only the linear relationship. Though the learning capacity may be increased by introducing the nonlinear functions, testing all the functions is impractical, and the function suitability to our problem is little known. Neural Network (NN), analogous to the biological nervous system, consists of input, hidden, and output layers comprising the neurons that identify the hidden pattern using the activation functions. The input layer receives the pre-processed data, whereas the output layer provides the results. It is the hidden layer that performs all the computations. Perceptron is the simplest and the oldest NN for performing linearly separable problems, generally binary classification. Figure 9.3 shows the basic artificial neuron structure. The frequently used predictor functions in artificial neural networks (ANN) are multilayer perceptron

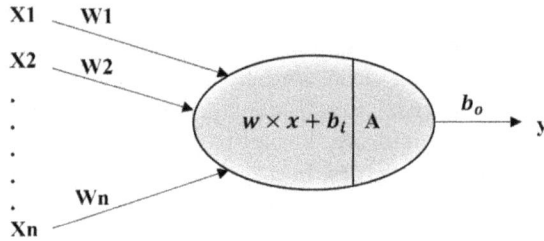

Figure 9.3 Basic artificial neuron structure.

(MLP) and radial basis function (RBF), which help in minimizing prediction errors. MLP is a fully connected feed-forward NN with backpropagation that adjusts the synaptic weights by backpropagating the error. The computational definition of an ANN is given in Equation 9.4.

$$y = A\left(w \times x + b_i\right) + b_o \tag{9.4}$$

where y and x denote the output and input vector, respectively, w is the weight matrix and b_i, b_o are the biases column vector from the input neurons to the hidden neurons and from the hidden neurons to the output neurons (Sharma et al., 2022).

ANN has been well exploited for classification and prediction tasks in PMCs. Convolutional neural network (CNN) commonly applied for image recognition consists of convolution layers that help in reducing the weights needed in the ML model resulting in efficient computation. Athijayamani et al. (2010) predicted the thrust force and torque during the drilling of roselle/sisal hybrid polyester composites based on the drilling process parameters using the ANN and nonlinear regression model. The predicted values showed that ANN resulted in fewer errors than the regression model. Prasad et al. (2016) utilized ANN to predict the flexural properties of coir/polyester composites with the input parameters of fiber length, fiber volume fraction, and specimen thickness.

9.5 PERFORMANCE ASSESSMENT METRICS

The prediction performance parameters are assessed to find the outperforming algorithm on the test data. There are dozens of evaluation metrics, but a few widely used will be discussed here. For classification tasks, a confusion matrix is used to determine various measures such as accuracy, precision, recall, and F-score. The commonly employed metrics for evaluating regression models' performance comprise MSE, MAE, RMSE, and R^2 (Botchkarev, 2019; Sokolova & Lapalme, 2009). Altarazi et al. (2019) studied the effect of film component proportions and processing parameters by predicting and classifying polymeric films' tensile properties using various ML algorithms. The specimens were fabricated by the two film production processes (compression and blow molding). The classification correctness was checked by the average accuracy, precision, and recall, whereas MAE and R^2 measured the prediction accuracy.

9.6 METRICS FOR CLASSIFICATION

The confusion matrix ($n \times n$ matrix) indicates the true and false prediction outcomes for a given test dataset, where n denotes the unique prediction classes. True positives (TP) refer to the number of instances the actual and the predicted classes are positive, whereas true negatives (TN) occur when both are negative. False positives (FP) are when the classifier misclassifies negatives as positives. False negatives (FN) are the vice versa of FP. This binary classification matrix can be expanded for multiclass classification. The diagonals of the multiclass classification matrix show the instances of true classification, and rest instances are misclassified. In this, TP, TN, FP, and FN are determined for each individual class (Sokolova & Lapalme, 2009). The columns of the confusion matrix represent the actual class, whereas the rows show the predicted class (Figure 9.4).

The few standard empirical measures from the confusion matrix are as follows.

9.6.1 Average accuracy

It is the fraction of instances correctly classified by the classifier (Equation 9.5)

$$Accuracy = (TP + TN) / (TP + TN + FP + FN) \tag{9.5}$$

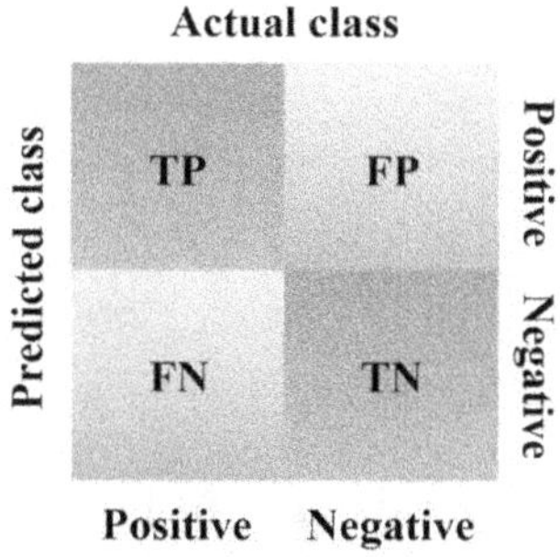

Figure 9.4 Confusion matrix for classification model.

9.6.2 Error rate

It depicts the fraction of incorrect predictions. It is also termed a misclassification rate, or classification error determined using Equation 9.6.

$$Error\ rate = (FP + FN) / (TP + TN + FP + FN) \qquad (9.6)$$

9.6.3 Precision

It indicates the fraction of true positive predictions in the positive prediction class (Equation 9.7).

$$Precision = TP / (TP + FP) \qquad (9.7)$$

9.6.4 Recall

It indicates the fraction of all positive samples predicted correctly as positive and also termed the True Positive Rate (TPR), sensitivity, and probability of detection. Equation 9.8 is used to compute the recall.

$$Recall = TP / (TP + FN) \qquad (9.8)$$

9.6.5 F-score

It is the harmonic mean of precision and recall as shown in Equation 9.9. For multiclass classification, the resulting F-score is the average of the individual class F-scores.

$$F - score = 2TP / (2TP + FP + FN) \qquad (9.9)$$

9.7 METRICS FOR REGRESSION

The regression model's performance metric determines how near or far the anticipated outcome is to the real one (Botchkarev, 2019). The most common metrics are as follows.

9.7.1 Mean absolute error

MAE is the computation of the average of absolute errors countered for the predicted values of the test dataset. In Equation 9.10, n represents the number of data points in the test set, A_i and P_i denote the actual value and predicted value, respectively, for ith instance.

$$MAE = \frac{1}{n} \sum_{i=1}^{n} \left| A_i - P_i \right| \tag{9.10}$$

9.7.2 Mean squared error/root mean square error

MSE/RMSE is the absolute gauge of the goodness of fit. MSE is the average of the squares of prediction errors, which depicts the spread of data around the regression line (Equation 9.11). It penalizes the outliers most, thus resulting in larger values. RMSE is the square root of MSE and is often used more than MSE. MSE values may sometimes be too large to handle. RMSE scales down the MSE values comparable to prediction errors for easy interpretation.

$$MSE = \frac{1}{n} \sum_{i=1}^{n} \left(A_i - P_i \right)^2 \tag{9.11}$$

9.7.3 Coefficient of determination

R^2 is the relative measure of how well the model fits the dependent variables. It is also termed the goodness of fit. It does not consider the overfitting problem. It is determined by Equation 9.12. SS_r and SS_m represent the squared sum error of the regression and mean line, respectively.

$$R^2 = 1 - \frac{SS_r}{SS_m} \tag{9.12}$$

9.8 CHALLENGES AND OPPORTUNITIES

ML has taken over the traditional methods of optimizing, predicting, classifying, and analyzing larger datasets and complex composite systems. Though ML is bringing a significant change in materials research by saving the costs and time incurred for investigation, the challenges encountered while handling the complexity need to be addressed. No algorithm performs ideally best for all types of problems. An algorithm may perform better in a specific problem only.

9.9 THE CURSE OF DIMENSIONALITY

The model performance depends not only on the algorithm deployed but on other factors also, such as the size and type of dataset, number of features, and data quality. Sufficient data with proper dimensionality is required to build a useful predictive model. One dimension corresponds to one feature/variable. Some algorithms may lose their reliability with increased input space dimensionality because of increased computational efforts for processing. The increased dimensionality requires a larger number of data points to evaluate the features' combination to develop an accurate model. This problem is referred to as the curse of dimensionality. It can be resolved by feature reduction techniques such as polynomial component analysis, regularization, and feature ranking. Unsupervised learning can help select the relevant features by visualizing the data. Polynomial component analysis has been used to eliminate the multicollinearity between the variables (Iguyon & Elisseeff, 2003; Mueller et al., 2016).

9.10 OVERFITTING/UNDERFITTING

Another challenge is overfitting/underfitting resulting from insufficient or low-quality data. Let's first understand the two terms before diving into the said challenge. Bias is the error rate of the training data. High bias means a high error rate. The other term is variance which denotes the error rate difference between training and testing data. Underfitting occurs when the model cannot capture the underlying trend in the dataset due to fewer data points or deploying the wrong algorithm. This type of model performs poorly on both the training and testing data. High bias and low variance cause underfitting, whereas overfitting occurs when the model is trained with massive data which captures the noise, resulting in poor performance for the new dataset (i.e., high variance). Low bias and high variance are the reasons for overfitting. The trade-off between bias and variance is necessary to avoid the problem. Cross-validation, regularization, and feature reduction practices can be helpful to avoid this issue, thus increasing the ML model efficiency (Sharma et al., 2022).

9.11 HANDLING MIXED VARIABLES

Due to the complicated system of PMCs, the datasets contain mixed datatype variables (continuous and discrete). It is challenging to analyze or draw conclusions from the mixed variables while using the clustering algorithms. The dataset can be handled using heterogeneous distance metrics, where the categorical and numerical data are treated differently to calculate the distance between the data points. Some algorithms cannot process the mixed variable for which the categorical data is encoded to the numerical form using techniques like label encoding, one hot transformation, and backward encoding. The way the variable is encoded also determines the quality of the model developed. This area concerning PMCs has not yet been exploited despite its highest need for optimum design (Mueller et al., 2016; Sharma et al., 2022).

Young researchers are exploiting the application of ML to the maximum extent and building hybrid machine learning to attain the combined benefits of individual algorithms. This includes the example of ensemble methods that play on how different models can be aggregated to attain the maximum prediction accuracy. Ensemble methods were utilized to predict the mechanical behavior of the PMCS. ML and PMCs have been extensively researched individually due to their innumerable benefits. But the execution of ML in materials science is still at a nascent stage. The substantial contrivance of ML can bring radical changes in materials science and discovery.

9.12 CONCLUSIONS

Soft computing techniques have leveraged research in the domain of science and engineering. This chapter has described basic supervised learning techniques used for different analyses in the field of PMCs. Majorly ML algorithms have been used to predict mechanical, tribological, and other properties as well as defect classification. The use of ML has the potential to concurrently achieve several adjustable multifunctional features in polymer composites depending on the constituents and their configurations. The ML model prediction performance metrics and their significance have been highlighted. Though ML outperforms conventional techniques, certain challenges need to be addressed and have been discussed. ML can be further exploited to articulate the selection of raw materials for the design and simulation system for numerable parameters leading to the manufacturing of optimal sustainable composite products.

REFERENCES

Alhijazi, M., Safaei, Babak, Zeeshan, Q., Asmael, Mohammed, Harb, Mohammad, & Qin, Z. (2022). An experimental and metamodeling approach to tensile properties of natural fibers composites. *Journal of Polymers and the Environment, 30*, 4377–4393. https://doi.org/10.1007/s10924-022-02514-1

Almanei, M., Oleghe, O., Jagtap, S., & Salonitis, K. (2021). Machine learning algorithms comparison for manufacturing applications. *Advances in Transdisciplinary Engineering, 15*, 377–382. https://doi.org/10.3233/ATDE210065

Altarazi, S., Allaf, R., & Alhindawi, F. (2019). Machine learning models for predicting and classifying the tensile strength of polymeric films fabricated via different production processes. *Materials, 12*(9). https://doi.org/10.3390/ma12091475

Amor, N., Noman, M. T., & Petru, M. (2021). Prediction of functional properties of nano TiO 2 coated cotton composites by artificial neural network. *Scientific Reports, 11*(1). https://doi.org/10.1038/S41598-021-91733-Y

Athijayamani, A., Natarajan, U., & Thiruchitrambalam, M. (2010). Prediction and comparison of thrust force and torque in drilling of natural fibre hybrid composite using regression and artificial neural network modelling. *International Journal of Machining and Machinability of Materials, 8*(1–2), 131–145. https://doi.org/10.1504/IJMMM.2010.034492

Botchkarev, A. (2019). A new typology design of performance metrics to measure errors in machine learning regression algorithms. *Interdisciplinary Journal of Information, Knowledge, and Management, 14*, 45–76. https://doi.org/10.28945/4184

Cao, R., Naya, S., Artiaga, R., García, A., & Varela, A. (2004). Logistic approach to polymer degradation in dynamic TGA. *Polymer Degradation and Stability, 85*(1), 667–674. https://doi.org/10.1016/J.POLYMDEGRADSTAB.2004.03.006

Chen, C. T., & Gu, G. X. (2019). Machine learning for composite materials. *MRS Communications, 9*(2), 556–566. https://doi.org/10.1557/mrc.2019.32

Choudhury, A. (2021). The role of machine learning algorithms in materials science: A state of art review on Industry 4.0. *Archives of Computational Methods in Engineering, 28*(5), 3361–3381. https://doi.org/10.1007/s11831-020-09503-4

Ciaburro, G., Puyana-Romero, V., Iannace, G., & Jaramillo-Cevallos, W. A. (2021). Characterization and modeling of corn stalk fibers tied with clay using support vector regression algorithms. *Journal of Natural Fibers*, 1–16. https://doi.org/10.1080/15440478.2021.1944427

Daghigh, V., Lacy, T. E., Daghigh, H., Gu, G., Baghaei, K. T., Horstemeyer, M. F., & Pittman, C. U. (2020). Machine learning predictions on fracture toughness of multiscale bio-nano-composites. *Journal of Reinforced Plastics and Composites, 39*(15–16), 587–598. https://doi.org/10.1177/0731684420915984

Dieterich, T. G. (2000). Ensemble methods in machine learning. *Lecture Notes in Computer Science (Including Subseries Lecture Notes in Artificial Intelligence and Lecture Notes in Bioinformatics), 1857 LNCS* (pp. 1–15). https://doi.org/10.1007/3-540-45014-9_1

Gairola, S., Naik, T. P., Sinha, S., & Singh, I. (2022). Corncob waste as a potential filler in biocomposites: A decision towards sustainability. *Composites Part C: Open Access, 9*, 100317. https://doi.org/10.1016/J.JCOMC.2022.100317

Gupta, P., Dwivedi, U. K., Yadav, V., & Kumar Yadav, A. (2022). Supervised classification model for estimation of wear in sisal fibre-epoxy composites. *Materials Today: Proceedings*. https://doi.org/10.1016/J.MATPR.2022.07.176

Iguyon, I., & Elisseeff, A. (2003). An introduction to variable and feature selection. *Journal of Machine Learning Research, 3*, 1157–1182.

Jayabal, S., Rajamuneeswaran, S., Ramprasath, R., & Balaji, N. S. (2013). Artificial neural network modeling of mechanical properties of calcium carbonate impregnated coir-polyester composites. *Transactions of the Indian Institute of Metals, 66*(3), 247–255. https://doi.org/10.1007/s12666-013-0255-9

Jiang, Z., Liang, Y., Su, Z., Chen, A., & Sun, J. (2021). Nondestructive testing of mechanical properties of bamboo–wood composite container floor by image processing. *Forests, 12*(11). https://doi.org/10.3390/f12111535

Kaushik, D., Gairola, S., Varikkadinmel, B., & Singh, I. (2022). Static and dynamic mechanical behavior of intra-hybrid jute/sisal and flax/kenaf reinforced polypropylene composites. *Polymer Composites*. https://doi.org/10.1002/PC.27114

Kazi, M. K., Eljack, F., & Mahdi, E. (2020a). Optimal filler content for cotton fiber/PP composite based on mechanical properties using artificial neural network. *Composite Structures*, *251*. https://doi.org/10.1016/J.COMPSTRUCT.2020.112654

Kazi, M. K., Eljack, F., & Mahdi, E. (2020b). Predictive ANN models for varying filler content for cotton fiber/PVC composites based on experimental load displacement curves. *Composite Structures*, *254*. https://doi.org/10.1016/J.COMPSTRUCT.2020.112885

Keerthi Gowda, B. S., Easwara Prasad, G. L., & Velmurugan, R. (2020). Prediction of mechanical strength attributes of coir/sisal polyester natural composites by ANN. *Journal of Soft Computing in Civil Engineering*, *4*(3), 79–105. https://doi.org/10.22115/SCCE.2020.226219.1200

Kibleur, P., Aelterman, J., Boone, M. N., Van den Bulcke, J., & Van Acker, J. (2022). Deep learning segmentation of wood fiber bundles in fiberboards. *Composites Science and Technology*, *221*. https://doi.org/10.1016/J.COMPSCITECH.2022.109287

Liao, S. H., Chu, P. H., & Hsiao, P. Y. (2012). Data mining techniques and applications - A decade review from 2000 to 2011. *Expert Systems with Applications*, *39*(12), 11303–11311. https://doi.org/10.1016/j.eswa.2012.02.063

Madhappan, R. K., Krishnasamy, S., & Alagirisamy, P. S. (2021). Prediction of hydration properties of rice straw mineral composite slabs using expert systems. *Journal of Thermoplastic Composite Materials*, *34*(6), 817–833. https://doi.org/10.1177/0892705719847246/FORMAT/EPUB

Mahajan, A., Binaz, V., Singh, I., & Arora, N. (2022). Selection of natural fiber for sustainable composites using hybrid multi criteria decision making techniques. *Composites Part C: Open Access*, *7*, 100224. https://doi.org/10.1016/j.jcomc.2021.100224

Mueller, T., Kusne, A. G., & Ramprasad, R. (2016). Machine learning in materials science: Recent progress and emerging applications. *Reviews in Computational Chemistry*, *29*(i), 186–273. https://doi.org/10.1002/9781119148739.ch4

Natural fibres show outstandingly low CO2 footprint compared to glass and mineral fibres - Renewable Carbon News. (n.d.). Retrieved September 7, 2022, from https://renewable-carbon.eu/news/natural-fibres-show-outstandingly-low-co2-footprint-compared-to-glass-and-mineral-fibres/

Nayak, S. K., & Satapathy, A. (2021). Mechanical and dry sliding wear characterization of marble dust and short kenaf fiber reinforced hybrid polyester composites. *Journal of Natural Fibers*. https://doi.org/10.1080/15440478.2021.1982815

Noryani, M., Sapuan, S. M., Mastura, M. T., Zuhri, M. Y. M., & Zainudin, E. S. (2018). A statistical framework for selecting natural fibre reinforced polymer composites based on regression model. *Fibers and Polymers*, *19*(5), 1039–1049. https://doi.org/10.1007/s12221-018-8113-3

Noryani, Muhammad, Sapuan, S. M., Mastura, M. T., Zuhri, M. Y. M., & Zainudin, E. S. (2019). Material selection of natural fibre using a stepwise regression model with error analysis. *Journal of Materials Research and Technology*, *8*(3), 2865–2879. https://doi.org/10.1016/j.jmrt.2019.02.019

Okuyama, M., Nakazawa, Y., & Funatsu, K. (2022). A data-driven scheme to search for alternative composite materials. *Science and Technology of Advanced Materials: Methods*, *2*(1), 106–118. https://doi.org/10.1080/27660400.2022.2063009

Osburg, V. S., Strack, M., & Toporowski, W. (2016). Consumer acceptance of wood-polymer composites: A conjoint analytical approach with a focus on innovative and environmentally concerned consumers. *Journal of Cleaner Production, 110,* 180–190. https://doi.org/10.1016/J.JCLEPRO.2015.04.086

Pattnaik, P., Sharma, A., Choudhary, M., Singh, V., Agarwal, P., & Kukshal, V. (2021). Role of machine learning in the field of Fiber reinforced polymer composites: A preliminary discussion. *Materials Today: Proceedings, 44,* 4703–4708. https://doi.org/10.1016/j.matpr.2020.11.026

Pereira Dias, P., Bhagya Jayasinghe, L., & Waldmann, D. (2021). Machine learning in mix design of Miscanthus lightweight concrete. *Construction and Building Materials, 302,* 124191. https://doi.org/10.1016/J.CONBUILDMAT.2021.124191

Pervaiz, M., & Sain, M. M. (2003). Carbon storage potential in natural fiber composites. *Resources, Conservation and Recycling, 39*(4), 325–340. https://doi.org/10.1016/S0921-3449(02)00173-8

Prabu, V. A., Kumaran, S. T., Uthayakumar, M., & Manikandan, V. (2018). Influence of redmud particle hybridization in banana/sisal and sisal/glass composites. *Particulate Science and Technology, 36*(4), 402–407. https://doi.org/10.1080/02726351.2016.1267284

Prasad, G. L. E., Gowda, B. S. K., & Velmurugan, R. (2016). Prediction of flexural properties of coir polyester composites by ANN. *Conference Proceedings of the Society for Experimental Mechanics Series, 7,* 173–180. https://doi.org/10.1007/978-3-319-21762-8_21

Probst, P., Wright, M. N., & Boulesteix, A. L. (2019). Hyperparameters and tuning strategies for random forest. *Wiley Interdisciplinary Reviews: Data Mining and Knowledge Discovery, 9*(3). https://doi.org/10.1002/WIDM.1301

Quinlan, J. R. (1986). Induction of decision trees. *Machine Learning, 1*(1), 81–106. https://doi.org/10.1023/A:1022643204877

Rajiv, B., Kalos, P., Pantawane, P., Chougule, V., & Chavan, V. (2022). Classification of damages in composite material using multi-support vector machine. *Journal of The Institution of Engineers (India): Series C, 103*(4), 655–661. https://doi.org/10.1007/S40032-022-00811-1

Ramesh, M., Palanikumar, K., & Reddy, K. H. (2017). Plant fibre based bio-composites: Sustainable and renewable green materials. *Renewable and Sustainable Energy Reviews, 79*(April 2016), 558–584. https://doi.org/10.1016/j.rser.2017.05.094

Rath, R. C., Baral, S. K., Singh, T., & Goel, R. (2022). Role of artificial intelligence and machine learning in product design and manufacturing. *2022 International Mobile and Embedded Technology Conference, MECON 2022,* 571–575. https://doi.org/10.1109/MECON53876.2022.9752455

Reitermanová, Z. (2010). Data splitting. *Week of Doctoral Students 2010 -- Proceedings of Contributed Papers,* 31–36.

Sharma, A., Mukhopadhyay, T., Rangappa, S. M., Siengchin, S., & Kushvaha, V. (2022). Advances in computational intelligence of polymer composite materials: machine learning assisted modeling, analysis and design. In *Archives of Computational Methods in Engineering* (Issue 0123456789). Springer Netherlands. https://doi.org/10.1007/s11831-021-09700-9

Sokolova, M., & Lapalme, G. (2009). A systematic analysis of performance measures for classification tasks. *Information Processing and Management, 45*(4), 427–437. https://doi.org/10.1016/j.ipm.2009.03.002

Su, Z., Jiang, Z., Liang, Y., Wang, B., & Sun, J. (2022). An artificial neural network model for predicting mechanical strength of bamboo-wood composite based on layups configuration. *BioResources, 17*(2), 3265–3277. https://doi.org/10.15376/biores.17.2.3265-3277

Sultana, N., Zakir Hossain, S. M., Alam, M. S., Islam, M. S., & Al Abtah, M. A. (2020). Soft computing approaches for comparative prediction of the mechanical properties of jute fiber reinforced concrete. *Advances in Engineering Software, 149.* https://doi.org/10.1016/J.ADVENGSOFT.2020.102887

Sun, X., He, M., & Li, Z. (2020). Novel engineered wood and bamboo composites for structural applications: State-of-art of manufacturing technology and mechanical performance evaluation. *Construction and Building Materials, 249.* https://doi.org/10.1016/J.CONBUILDMAT.2020.118751

Tholkappiyana, E., Saravanan, D., Jagasthitha, R., Angeswari, T., & Surya, V. T. (2015). Modelling of sound absorption properties of sisal fibre reinforced paper pulp composites using regression model. *Indian Journal of Fibre and Textile Research, 40*(1), 19–24.

Vigneshwaran, K., & Venkateshwaran, N. (2019). Statistical analysis of mechanical properties of wood-PLA composites prepared via additive manufacturing. *International Journal of Polymer Analysis and Characterization, 24*(7), 584–596. https://doi.org/10.1080/1023666X.2019.1630940

Wang, Z., Chegdani, F., Yalamarti, N., Takabi, B., Tai, B., El Mansori, M., & Bukkapatnam, S. (2020). Acoustic emission characterization of natural fiber reinforced plastic composite machining using a random forest machine learning model. *Journal of Manufacturing Science and Engineering, Transactions of the ASME, 142*(3). https://doi.org/10.1115/1.4045945

You, G., Wang, B., Li, J., Chen, A., & Sun, J. (2022). The prediction of MOE of bamboo-wood composites by ANN models based on the non-destructive vibration testing. *Journal of Building Engineering, 59.* https://doi.org/10.1016/J.JOBE.2022.105078

Zeng, G. S., Hu, C., Zou, S., Zhang, L., & Sun, G. (2019). BP neural network model for predicting the mechanical performance of a foamed wood-fiber reinforced thermoplastic starch composite. *Polymer Composites, 40*(10), 3923–3928. https://doi.org/10.1002/pc.25252

Chapter 10

Explainable machine learning model for Industrial 4.0

Ajay Kumar Badhan
Lovely Professional University, Phagwara, India

Rana Gill
Chandigarh University, Mohali, India

10.1 INTRODUCTION

Industry 4.0 is a term used to describe the Fourth Industrial Revolution that is currently underway. It represents a significant shift in how industries operate and is characterized by integrating digital technologies, automation, and the Internet of Things (IoT) into manufacturing and other industries. The Industry 4.0 revolution promises to bring about a new era of productivity, efficiency, and innovation, but it also presents significant challenges for businesses and workers (Farooq & Waseem, 2020).

One of the key features of Industry 4.0 is the use of smart systems, which can communicate with each other and humans in real time. These systems rely on data collected from sensors, machines, and other devices to make decisions and optimize processes. This creates a highly connected and responsive ecosystem that can adapt quickly to changing conditions. Another important aspect of Industry 4.0 is automating production processes and increasing efficiency. This includes using robots, drones, and autonomous vehicles, which can perform tasks more quickly and accurately than humans.

The industry's evolution from 1.0 to 4.0 has been marked by significant technological advancements transforming how goods are produced and consumed. Each industrialization phase has been characterized by a major shift in industries' operations, with new technologies and innovations driving productivity, efficiency, and growth. The diagrammatic view of the evolution of industries is presented in Figure 10.1.

- First Industrial Revolution
  It began with the mechanization of production through water and steam power (Xu et al., 2018). This allowed for the mass production of goods and marked a significant shift away from manual labor.
- Second Industrial Revolution
  This era was marked by the introduction of electricity, which powered new inventions such as the assembly line and the conveyor belt

DOI: 10.1201/9781003453567-10

The Evolution of Industry
Ages

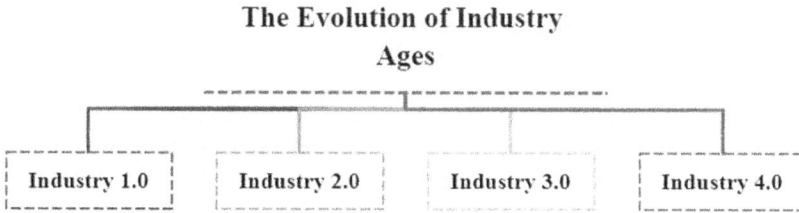

Figure 10.1 Evolution of the industrial ages.

(Alaloul et al., 2019). This allowed for even greater productivity and efficiency in manufacturing.

- Third Industrial Revolution
  It was the introduction of computers and automation, enabling greater control over production processes and the development of more complex products (Wächter, 2020). This industrialization phase also paved the way for the digital age with the growth of the internet and the proliferation of mobile devices.

- Fourth Industrial Revolution
  It is the current phase of industrialization that is underway. It is characterized by integrating digital technologies, automation, and the Internet of Things (IoT) into manufacturing and other industries (Xu et al., 2018). This includes using smart systems that can communicate with each other and humans in real time using robotics, artificial intelligence, and advanced analytics to optimize production processes.

10.2 EXPLAINABLE MACHINE LEARNING: TECHNIQUES FOR INDUSTRY 4.0

Explainable machine learning (XAI) is a subfield of artificial intelligence (AI) that focuses on making ML models more interpretable and transparent. XAI aims to provide insights into how a particular model makes its decisions and ensure they are fair, ethical, and accountable (Duarte, 2020). The techniques that are implemented for creating ML models are classified as follows.

10.2.1 Rule-based systems

These systems are designed to mimic the decision-making process of human experts by using a set of rules to conclude. Rule-based systems are easy to interpret and explain because the decision-making process is transparent (Kumar & Sinha, 2019). The basic structure of the rule-based system is shown in Figure 10.2.

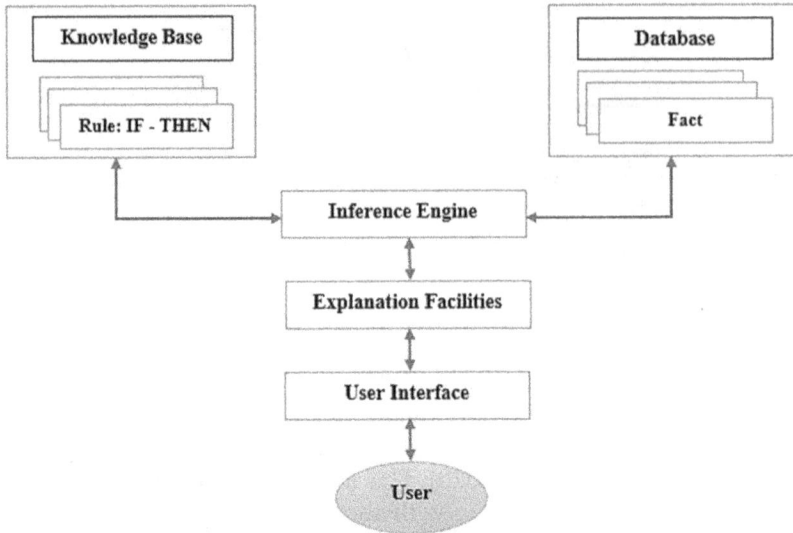

Figure 10.2 Basic structure of the rule-based system.

10.2.2 Decision trees

Decision trees are a popular technique for creating explainable ML models. They work by recursively partitioning the data into subsets based on the values of the input variables, and they generate a tree-like structure that represents the decision-making process (Duarte, 2020). The tree branches represent the decision rules, and the leaves represent the final decision. The basic diagram for the decision tree to determine whether a person is fit or unfit is presented as follows.

10.2.3 Linear models

Linear models like linear and logistic regression are simple and interpretable ML models. They work by fitting a linear equation to the data, which can be easily interpreted and explained (Piekutowska et al., 2021).

10.2.4 Local interpretable model – agnostic explanations (Lime)

LIME is a technique that provides interpretable explanations of individual predictions made by a black box model (Zafar, 2021). LIME works by generating a simpler, more interpretable model that approximates the behavior of the black box model for a specific input.

10.2.5 Shapely additive explanations (SHAP)

SHAP is a framework for explaining the output of any ML model. It was introduced in a 2017 paper by Lundberg and Lee and has become increasingly popular in the ML community. SHAP provides a way to assign a score to each feature of an input, indicating how much that feature contributes to the model's output. This score is based on the concept of Shapley values from cooperative game theory, which measures the marginal contribution of a feature to a group of features.

The SHAP framework can be used for various tasks like feature selection, model debugging, and model interpretation. It can also be applied to various models, including neural networks, decision trees, and linear models. One of the strengths of SHAP is that it can handle complex interactions between features. For example, if two features are highly correlated, SHAP will assign scores that reflect the joint effect of both features, rather than just the effect of each feature in isolation. SHAP is a powerful tool for understanding how ML models make predictions and gaining insights into the underlying data (Mangalathu et al., 2020).

10.3 METHODOLOGY EXPLANATION

10.3.1 Rule-based systems

A rule-based system is a methodology used in ML that involves creating a set of rules or logical statements that allow the system to make decisions or predictions based on input data. In the context of Industry 4.0, a rule-based system can be used to create an explainable ML model that can be used to make decisions or predictions in real time. A rule-based system for machine learning is shown in Figure 10.3.

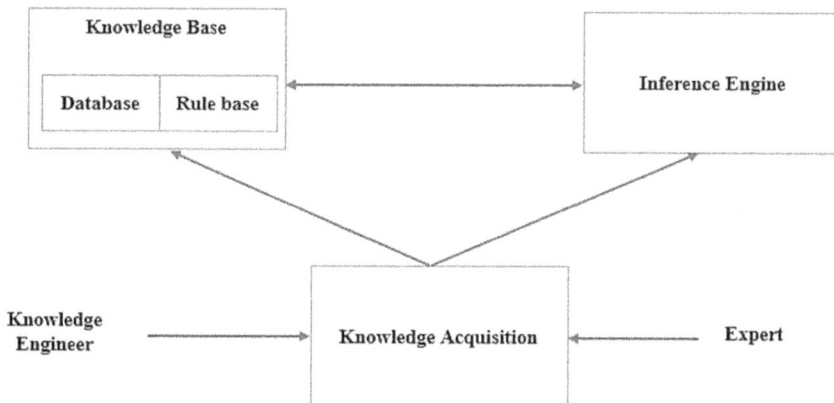

Figure 10.3 A rule-based system for machine learning.

The methodology for building a rule-based system typically involves several steps:

- Define the problem: The first step is clearly defining the problem you are trying to solve. This involves identifying the input data used to make decisions or predictions and the output the system should produce.
- Identify the rules: Once you have defined the problem, you need to identify the rules or logical statements that will allow the system to make decisions or predictions based on the input data. These rules can be based on domain expertise, existing knowledge, or data-driven insights.
- Implement the rules: After identifying the rules, you need to implement them so that the system can use them to make decisions or predictions in real time. This may involve writing code or configuring the system to apply the rules to incoming data.
- Test and refine: Once the rules are implemented, you must test the system to ensure it produces the desired output. If the system is not producing the desired output, you may need to refine the rules or adjust the input data to improve the system's performance.

Different rule-based systems are implemented depending on the specific problem we are trying to solve and the data type we are working with. Here are some of the most common types of rule-based systems:

- Expert systems: Expert systems are rule-based systems designed to replicate human experts' decision-making processes. These systems typically use a set of rules and a knowledge base to make decisions or predictions based on input data (Linko, 1998).

  Example: An example of an expert system might be a medical diagnosis system that uses a set of rules and a knowledge base to diagnose patients with various medical conditions. The system might ask users questions about their symptoms, medical history, and other factors and then use rules to identify the most likely diagnosis.
- Fuzzy logic systems: Fuzzy logic systems are rule-based systems that can handle uncertain or imprecise data. Unlike traditional Boolean logic systems, which rely on binary true/false statements, fuzzy logic systems use a degree of membership to represent the degree to which an input belongs to a particular category (Duarte, 2020).

  Example: A common example of a fuzzy logic system is a thermostat that adjusts the temperature in a room based on the current temperature and the desired temperature. Rather than using a simple binary true/false statement to determine whether the temperature is too high or too low, the system might use a degree of membership to represent the degree to which the current temperature is too high or too low.

10.3.2 Decision trees

Decision trees are rule-based systems that use a tree-like structure to represent a series of decisions or predictions. Each node in the tree represents a decision or prediction based on a particular input variable, and the tree branches represent the possible outcomes or decisions that can be made (Yeo & Grant, 2018).

Example: An example of a decision tree might be a credit risk assessment system that uses a series of questions to determine a loan applicant's creditworthiness. The system might start by asking whether the applicant has a steady source of income and then use the answer to that question to determine the next question to ask (e.g., if the answer is "yes," the system might ask about the applicant's credit score; if the answer is "no," the system might ask about the applicant's employment history).

- Production systems: Production systems are rule-based systems that automate complex decision-making processes. These systems typically consist of rules and a control system that executes the rules in response to incoming data (Yin et al., 2017).

  Example: An example of a production system might be a quality control system in a manufacturing plant. The system might use a set of rules and sensors to detect product defects as they are being manufactured and then use a control system to adjust the manufacturing process to eliminate them automatically.

10.3.3 Rule-based classifiers

Rule-based classifiers are a type of rule-based system that is used to classify incoming data into different categories. These systems typically use a set of rules and a decision-making algorithm to determine the most likely category for a given input (Peter et al., 2008).

Example: An example of a rule-based classifier might be a spam filter that uses a set of rules to determine whether an incoming email is spam. The system might use rules to analyze the content of the email (e.g., if the email contains certain keywords, it is more likely to be spam) as well as other factors (e.g., if the email is from an unknown sender, it is more likely to be spam).

10.3.4 Decision tree

It is a popular and widely used method in explainable ML models in Industry 4.0. It is a predictive model with a tree-like structure to map decisions and other possible consequences. The tree is built by recursively splitting the data into subsets based on the features that provide the most information gained until a leaf node that provides a decision or prediction is reached. For example, consider determining whether the person is fit or unfit based on BMI and age. The diagrammatic view is presented in Figure 10.4.

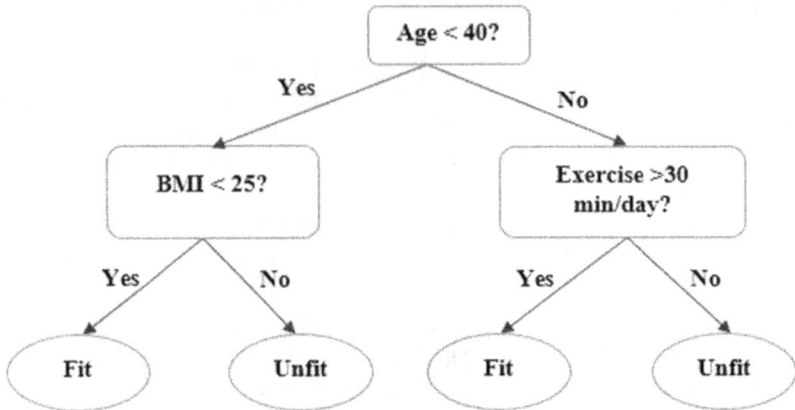

Figure 10.4 Decision tree diagram for determining whether a person is fit or unfit.

In the above example, the first decision node determines whether the person's age is less than 40. If yes, the tree moves to the left branch and determines whether their BMI (body mass index) is less than 25. If yes, the person is classified as fit (represented by the leaf node on the leftmost branch). If the person's BMI is greater than or equal to 25, the tree moves to the right branch and asks whether they exercise for more than 30 minutes daily. If the answer is yes, then the person is classified as fit. If the answer is no, then they are classified as not fit.

In Industry 4.0, decision trees can be used in various applications, such as predicting equipment failure, optimizing production processes, and identifying quality control issues. Decision trees are beneficial when many variables can impact the outcome, and the relationships between the variables are complex. The parameters that are considered for building decision trees are as follows:

- Splitting criteria: The criteria used to determine the best way to split the data at each node in the tree (Charbuty & Abdulazeez, 2021). Common criteria include information gain, Gini impurity, and Entropy.
- Maximum depth: The maximum number of levels in the decision tree. This parameter prevents overfitting and ensures the tree is not too complex.
- Minimum sample split: The minimum number of samples required to split a node. This parameter is used to prevent the tree from being split too many times and to avoid overfitting.
- Minimum sample leaf: The minimum number of samples required at a leaf node. This parameter is used to prevent the tree from being too specific and to ensure that the tree is not too complex.

- The maximum number of features: The maximum number of features to consider when searching for the best split at each node. This parameter is used to reduce the complexity of the decision tree and to improve its efficiency.
- Pruning: Removing branches from the decision tree that do not improve its performance (Lu & Ma, 2020). This is typically done using a validation set or cross-validation.

In decision tree models, the *entropy function* is often used as a criterion function to measure the information gained from each feature in splitting the data. Entropy measures the level of uncertainty or randomness in a set of data, and the information gain is the difference between the original dataset's Entropy and the resulting dataset's weighted average Entropy after splitting (Sethi, 1990). By maximizing the information gain, the decision tree algorithm can identify the best features to split on and create a tree that minimizes the overall Entropy of the data, making it a useful tool for classification and prediction tasks. The formula for Entropy is given is shown in Equation 10.1.

$$Entropy = -p_1 log_2(p_1) - p_2 log_2(p_2) - \ldots - p_n * log_2(p_n) \tag{10.1}$$

where $p_1, p_2, \ldots\ldots p_n$ are the proportions of samples in the node that belong to each class. The logarithm base 2 ensures that the Entropy is measured in bits (Mahbooba et al. 2021).

- Information gain measures the reduction in Entropy achieved by splitting the data based on a particular feature. The idea is to choose the feature that results in the largest reduction in Entropy since this feature is likely to be the most informative for classifying the samples. The formula for information gain is shown in Equation 10.2:

$$IG = H(P) - \sum \left(\left(|C_i| |P| \right) * H(C_i) \right) \tag{10.2}$$

where

- IG: Information Gain
- $H(P)$: Entropy of the parent node
- Σ: Summation symbol
- $|C_i|$: Number of instances in the ith child node
- $|P|$: Total number of instances in the parent node
- $H(C_i)$: Entropy of the ith child node resulting from the split.

The weighted average is calculated based on the proportion of samples that belong to each child node (Charbuty & Abdulazeez, 2021).

Example: Suppose we have a dataset with 100 samples, where 60 samples belong to class A and 40 belong to class B. If we split the data based on a certain feature and obtain two child nodes with 70 samples and 30 samples, respectively, where 50 samples in the first node belong to class A and 20 samples in the second node belong to class B, then the Entropy of the parent node is:

$$\text{Entropy (Parent)} \left[H(P) \right] = -(60/100) \times log2(60/100)$$
$$- (40/100) \times log2(40/100)$$
$$= 0.971$$

And the Entropy of the child nodes is:

$$\text{Entropy(child)} \left[H(C_i) \right] = [70/100] \times \left(-(50/70) \times log2(50/70) \right.$$
$$- (20/70) log2(20/70))$$
$$+ [30/100] - ((10/30)$$
$$\times log2(10/30) - (20/30) \times log2(20/30))$$
$$= 0.823$$

So, the information gained is:

$$\text{Information Gain} (IG) = 0.971 - [70/100] \times 0.823 = 0.341$$

This means that splitting the data based on this feature resulted in a reduction in Entropy of 0.341 bits, which is the amount of information gained by the split.

10.3.5 Local interpretable model – agnostic explanations (LIME)

It is an explainability technique that explains the predictions of ML models at the local level. It can be used to interpret the decisions made by a black box ML model by approximating it with a simpler, more transparent model that is easier to understand (Ribeiro & Guestrin, 2016).

In the context of Industry 4.0, LIME can provide transparency and interpretability to the decision-making process of ML models used in industrial settings. The following steps can be taken to implement a LIME-based explainability model for Industry 4.0:

- Train a machine learning model: The first step is to train an ML model on the data collected from the industrial processes. This model can be a black box model, such as a deep neural network or a decision tree.

- Select a sample to explain: Once the model is trained, select a sample from the dataset you want to explain. This sample should be representative of the dataset as a whole.
- Generate local explanations: Next, generate local explanations for the selected sample using LIME. This involves approximating the black box model with a simpler, more transparent model that can be easily understood. The local explanations are generated by perturbing the input features of the selected sample and observing the resulting changes in the model's output. These perturbations are then used to fit a simpler model that can be used to explain the predictions of the black box model for the selected sample.
- Visualize the explanations: Finally, visualize the local explanations using plots, graphs, or other visualization tools. This helps to understand better how the model makes decisions and what factors contribute to those decisions.

  Example: Consider a scenario where a manufacturing company is using an ML model to predict the quality of their products based on several input features such as temperature, humidity, pressure, and time. The model is a black box model, and the company wants to understand how the model makes predictions and which input features are most important. The LIME-based explain ability model can be used in this scenario as follows:

  - Train a machine learning model: The first step is to train an ML model using the historical data collected from the manufacturing processes. Let's say that the company uses a random forest model to predict product quality.
  - Select a sample to explain: Once the model is trained, the company selects a sample from the dataset they want to explain. Let's say that the selected sample is a product manufactured at a temperature of 200 degrees, a humidity of 60%, a pressure of 100 kPa, and a time of 30 minutes.
  - Generate local explanations: Next, the company generates local explanations for the selected sample using LIME. LIME works by perturbing the input features of the selected sample and observing the resulting changes in the model's output. Let's say that the company perturbs the temperature feature and generates a set of perturbations around the temperature of the selected sample. The resulting changes in the predicted quality of the product are then used to fit a simpler model that can be used to explain the predictions of the random forest model. The simpler model might be a linear regression model showing how the temperature feature contributes to the predicted product quality.
  - Visualize the explanations: Finally, the company visualizes the local explanations using plots or graphs. For example, they might plot how the predicted product quality changes as the temperature

feature is perturbed around the selected sample's temperature. This can help the company understand how the random forest model makes predictions and which input features are most important for predicting product quality (Gabbay et al., 2021).

10.3.6 Shapely additive explanations (SHAP)

It is a framework for model interpretation and feature attribution in ML. It aims to explain individual predictions made by a model by decomposing the prediction into contributions from each feature. This allows for a better understanding of the model's decisions, which can be useful for various applications, including in Industry 4.0 (Rathi, 2018).

It builds on the idea of additive feature attribution, which means that the contribution of each feature to the prediction is measured by how much the feature's value deviates from a baseline value. The baseline value is usually the mean value of the feature in the training data. By comparing the actual feature value to the baseline value, SHAP can calculate the effect of each feature on the prediction.

Considering a scenario for predicting the likelihood of the customer churning (i.e., canceling the subscription or service) based on customer attributes such as age, gender, location, and purchase history, we develop an ML model. Utilizing the SHAP, one can explain why a model made a particular prediction for a specific customer.

Example: Suppose the model predicted Customer A's likelihood of churning. We can use SHAP to decompose this prediction into the contributions from each customer attribute. The SHAP process will be as follows:

- Select a specific prediction from the model, in this case, the prediction for Customer A.
- Create a baseline dataset that represents the average customer's attributes. For example, we might use the mean values for age, gender, location, and purchase history.
- Calculate the SHAP values for each attribute for the prediction of Customer A. These values represent the contribution of each attribute to the final prediction relative to the baseline dataset.
- Visualize the SHAP values in a way that is easy to interpret. For example, one can create a bar chart that shows the magnitude and direction of each attribute's contribution to the prediction.

10.4 LINEAR MODELS

Linear models are ML models that use a linear function to predict the output variable based on one or more input variables. They are commonly used for regression and classification tasks. In a linear regression model, the output variable is a continuous value, and the model predicts this value based on a set of input variables. The model's prediction is based on a linear combination of the input variables, each weighted by a coefficient. The coefficients are learned during training, and the goal is to minimize the difference between the predicted and true values in the training data (Chinnalagu & Durairaj, 2021).

In a linear classification model, the output variable is a discrete value, and the model predicts this value based on a set of input variables. The model's prediction is based on a linear combination of the input variables, each weighted by a coefficient. However, instead of predicting the output directly, the model calculates the probability of each class and selects the class with the highest probability as the output.

There are several types of linear models, including the following:

- Ordinary least squares (OLS): A linear regression model that minimizes the sum of the squared differences between the predicted and true values
- Ridge regression: A linear regression model that uses L2 regularization to prevent overfitting
- Lasso regression: A linear regression model that uses L1 regularization to encourage sparsity in the coefficient values
- Logistic regression: A linear classification model that uses the logistic function to calculate the probability of each class
- Linear support vector machines (SVM): A linear classification model that separates the classes with a hyperplane and maximizes the margin between the classes

Linear models are popular because they are simple, interpretable, and computationally efficient. However, they may not perform as well as more complex models for some tasks, especially those with complex relationships between the input and output variables.

10.4.1 Examples of linear models and their applications

- Linear regression: A real estate agent wants to predict the selling price of a house based on its square footage. The agent collects data on recent home sales in the area, including each home's square footage and selling price. The agent can use a linear regression model to predict the selling price of a new house based on its square footage.
- Ridge regression: A pharmaceutical company wants to predict the effectiveness of a new drug based on its chemical properties. The company collects data on the chemical properties of existing drugs and their effectiveness. The company can use a ridge regression model to

predict the new drug's effectiveness, while preventing overfitting of the model to the training data.

- Lasso regression: A marketing team wants to predict which customers will most likely purchase a new product. The team collects data on customer demographics, purchasing history, and marketing campaigns. The team can use a lasso regression model to identify the most important features for predicting customer behavior and focus their marketing efforts accordingly.
- Logistic regression: A medical researcher wants to predict whether a patient has a particular disease based on their symptoms. The researcher collects data on patients who have been diagnosed with the disease and those who have not. The researcher can use a logistic regression model to calculate the patient's disease probability based on symptoms.
- Linear Support Vector Machines (SVM): A credit card company wants to predict which customers will most likely default on their payments. The company collects data on customer credit history, payment behavior, and other factors. The company can use a linear SVM model to separate the customers into two groups – those who are likely to default and those who are not – and identify the most important factors for predicting default risk (Chen et al., 2021).

Table 10.1 overviews some popular XAI techniques, their applications, and their benefits, and several popular XAI techniques are explained, including rule-based systems, decision trees, linear models, LIME, and SHAP. Each

Table 10.1 XAI techniques, applications, and benefits.

XAI Techniques	Applications	Benefits
Rule-based systems	Predictive maintenance, quality control	Easy to understand and interpret, can capture expert knowledge
Decision trees	Predictive maintenance, anomaly detection	Can handle non-linear relationships and interactions between variables
Linear models	Predictive maintenance, quality control	Easy to interpret, provide insights into the impact of different features
LIME	Anomaly detection	It can provide explanations for individual predictions, helpful in identifying data biases
SHAP	Quality control	Can provide feature importance rankings and insights into how different features impact model predictions

method has specific applications, benefits, and advantages. Rule-based systems are easy to understand and capture expert knowledge, decision trees can handle non-linear relationships, linear models are easy to interpret, LIME can explain individual predictions, and SHAP can provide feature importance rankings and insights into how different features impact model predictions.

10.5 IMPLICATIONS OF EXPLAINABLE MACHINE LEARNING FOR INDUSTRY 4.0

Integrating advanced technologies such as IoT, cloud computing, and ML has become more commonplace in industrial processes, and the need for transparency and interpretability in ML models has become increasingly important. XAI techniques such as rule-based systems, decision trees, linear models, LIME, and SHAP can ensure that ML models are fair, ethical, and accountable, providing transparency and interpretability essential for building trust in ML models.

One of the main implications of this chapter is that any organization looking to implement ML in its industrial processes should make XAI an essential consideration. By doing so, organizations can ensure that their models are reliable, trustworthy, and aligned with ethical and regulatory requirements. XAI techniques can also provide various benefits, such as identifying data biases, capturing expert knowledge, and understanding the impact of different features on model predictions.

Another important implication of this chapter is that XAI is critical for building trustworthy and reliable ML models in Industry 4.0. ML models that are transparent and interpretable can help to reduce the risk of errors and biases and can also improve decision-making processes in industrial applications. XAI can also help improve the accountability of ML models, allowing stakeholders to understand how decisions are being made and identify potential issues or errors.

Overall, the chapter on XAI for Industry 4.0 highlights the importance of transparency and interpretability in ML models and provides a range of techniques that can be used to achieve these goals. By adopting XAI techniques, organizations can ensure that their ML models are trustworthy, reliable, and aligned with ethical and regulatory requirements. They can also provide various benefits in identifying data biases, capturing expert knowledge, and improving decision-making processes.

10.6 CONCLUSIONS

In conclusion, XAI is essential for building trustworthy and reliable ML models in Industry 4.0. As the integration of advanced technologies like IoT, cloud computing, and ML becomes more commonplace in industrial processes, the need for transparency and interpretability in ML models has become increasingly important. In this chapter, we provided an overview of various XAI techniques, such as rule-based systems, decision trees, linear models, LIME, and SHAP, and demonstrated how each method could be used in Industry 4.0 applications. Our analysis shows that XAI techniques can ensure that ML models are fair, ethical, and accountable, providing transparency and interpretability essential for building trust in ML models. We conclude that any organization looking to implement ML in its industrial processes should consider XAI important to ensure its models are reliable, trustworthy, and aligned with ethical and regulatory requirements.

10.6.1 Future scope

The integration of advanced technologies such as IoT, cloud computing, and ML into industrial processes is expected to continue to grow rapidly in the future. As the adoption of ML in Industry 4.0 expands, there will be an increasing need for transparent and interpretable ML models to ensure that they are trustworthy and aligned with ethical and regulatory requirements. Future research in the field of XAI for Industry 4.0 should focus on developing new and improved techniques to understand better how ML models make predictions and decisions. This may involve the development of novel visualization methods, the integration of domain-specific knowledge into ML models, and the use of new model architectures that are designed to be more interpretable.

In addition, future research should also focus on addressing the challenges associated with deploying XAI techniques in real-world industrial applications. This may involve the development of new tools and frameworks that can help organizations to integrate XAI into their existing ML workflows, as well as the result of new evaluation metrics and techniques for assessing the performance and interpretability of XAI models.

Overall, the future scope of XAI for Industry 4.0 is broad and multifaceted. As the adoption of ML in industrial processes continues to grow, XAI will become an increasingly important tool for ensuring that ML models are transparent, interpretable, and aligned with ethical and regulatory requirements. Future research will enable organizations to build trustworthy and reliable ML models to drive innovation and growth in the Industry 4.0 era.

REFERENCES

Alaloul, W. S., Liew, M. S., Amila, N., Abdullah, W., & Kennedy, I. B. (2019). Industrial revolution 4. 0 in the construction industry: Challenges and opportunities for stakeholders. *Ain Shams Engineering Journal*, 1(11), 225–230.

Charbuty, B., & Abdulazeez, A. (2021). Classification based on decision tree algorithm for machine learning. *Journal of Applied Science and Technology Trends*, 2(01), 20–28. https://doi.org/10.38094/jastt20165

Chen, H., Zhang, C., Jia, N., Duncan, I., Yang, S., & Yang, Y. (2021). A machine learning model for predicting the minimum miscibility pressure of CO2 and crude oil system based on a support vector machine algorithm approach. *Fuel*, *290*, 120048. https://doi.org/10.1016/j.fuel.2020.120048

Chinnalagu, A., & Durairaj, A. K. (2021). Context-based sentiment analysis on customer reviews using machine learning linear models. *PeerJ Computer Science*, *7*, e813. https://doi.org/10.7717/peerj-cs.813

Duarte, M. F. (2020). Explainable machine learning for scientific insights and discoveries. *IEEE Access*, *8*(1), 42200–42216. https://doi.org/10.1109/ACCESS.2020.2976199

Farooq, M. U., & Waseem, M. (2020). A review on Internet of Things (IoT). *International Journal of Computer Applications*, *113*(1), 1–7.

Gabbay, F., Bar-Lev, S., Montano, O., & Hadad, N. (2021). A LIME-based explainable machine learning model for predicting the severity level of COVID-19 diagnosed patients. *Applied Sciences*, *11*(21), 10417. https://doi.org/10.3390/app112110417

Kumar, V., & Sinha, D. (2019). An integrated rule based intrusion detection system: Analysis on UNSW-NB15 data set and the real time online dataset. *Cluster Computing*, *0123456789*(1), 1–21. https://doi.org/10.1007/s10586-019-03008-x

Linko, S. (1998). Expert systemsÐ what can they do for the food industry? *Trends in Food Science & Technology*, *9*(1), 3–12.

Lu, H., & Ma, X. (2020). Hybrid decision tree-based machine learning models for short-term water quality prediction. *Chemosphere*, *249*, 126169. https://doi.org/10.1016/j.chemosphere.2020.126169

Mahbooba, B, Timilsina, M, Sahal, R, Serrano, M. 2021. Explainable artificial intelligence (XAI) to enhance trust management in intrusion detection systems using decision tree model. *Complexity*. 2021:1–11.

Mangalathu, S., Hwang, S., & Jeon, J. (2020). Failure mode and effects analysis of RC members based on machine-learning-based SHapley Additive exPlanations (SHAP) approach. *Engineering Structures*, *219*(February), 110927. https://doi.org/10.1016/j.engstruct.2020.110927

Peter, F., Gyenesei, A., & Abonyi, J. (2008). Compact fuzzy association rule-based classifier. *Expert Systems with Applications*, *34*(1), 2406–2416.

Piekutowska, M., Niedbała, G., Piskier, T., & Lenartowicz, T. (2021). The application of multiple linear regression and artificial neural network models for yield prediction of very early potato cultivars before harvest. *Agronomy*, *11*(4), 885.

Rathi, S. (2018). Generating counterfactual and contrastive explanations using SHAP. *ArXiv Preprint ArXiv*, *10*(1), 1–6.

Ribeiro, M. T., & Guestrin, C. (2016). Model-agnostic interpretability of machine learning. *ArXiv Preprint ArXiv*, *1*(Whi), 1–5.

Sethi, I. K. (1990). Entropy nets: From decision trees to neural networks. *Proceedings of the IEEE*, *78*(10), 1605–1613.

Wächter, C. (2020). Challenging ecodesign. *LED Professional*, 1–8.

Xu, M., David, J. M., & Kim, S. H. (2018). The fourth industrial revolution: Opportunities and challenges. *International Journal of Financial Research*, *9*(2), 90–95.

Yeo, B., & Grant, D. (2018). Predicting service industry performance using decision tree analysis. *International Journal of Information Management*, *38*(October 2017), 2017–2019.

Yin, Y., Stecke, K. E., & Li, D. (2017). The evolution of production systems from Industry 2.0 through Industry 4.0. *International Journal of Production Research*, *7543*(November), 1–14.

Zafar, M. R. (2021). Deterministic local interpretable model-agnostic explanations for stable explainability. *Machine Learning and Knowledge Extraction*, *3*(3), 525–541.

Chapter 11

Applications of machine learning in the manufacturing sector

Concept, framework, and scope

Pankaj Sawdatkar, Dilpreet Singh, and Krishnendu Kundu

Academy of Scientific and Innovative Research, (AcSIR),
Ghaziabad, India

CSIR – Central Mechanical Engineering and Research Institute –
CoEFM, Ludhiana, India

11.1 INTRODUCTION

Manufacturing is one of the prominent sectors of the world economy. It is the main source of GDP, approximately 16% of the global GDP in 2019 (Rai et al., 2021). Manufacturing is defined as converting raw material into some useful product with the help of a fabrication technique and assembly of components. Since the industrial revolution started, making products in large numbers with less cost and high quality has been the main critical point in front of the manufacturing sector. Industrial revaluation started in the 18th century in England, and from that point to date manufacturing industry has changed dramatically over the years. Many improvements happened in the manufacturing sector, and these improvements are classified according to time zone, and the particular development happened in that era as the industrial revolution started from Industry 1.0 and reached Industry 4.0. The advancement of technology increases the use of computers in manufacturing as more machines are used, which are controlled by computers. These machines are synchronized to achieve a task to save time and human effort and increase productivity. Now, the world is shifted to Industry 4.0, called the Fourth Industrial Revolution, which makes manufacturing smart with the help of advanced computing power. This computing power connects the machine; therefore, various machine learning (ML) and (artificial intelligence (AI) algorithms make the system intelligent. Flexible automation makes manufacturing more sustainable and accurate with higher quality and maximizes efficiency and productivity (Rai et al., 2021).

The evolution of the manufacturing industry can be divided into four industrial revolutions. Figure 11.1 shows the schematic of these four industrial revolutions. Industry 1.0 started in the late 18th century, with mechanical production using steam power. Industry 2.0 followed in the early 20th century, with the introduction of mass production using electricity. Industry

DOI: 10.1201/9781003453567-11

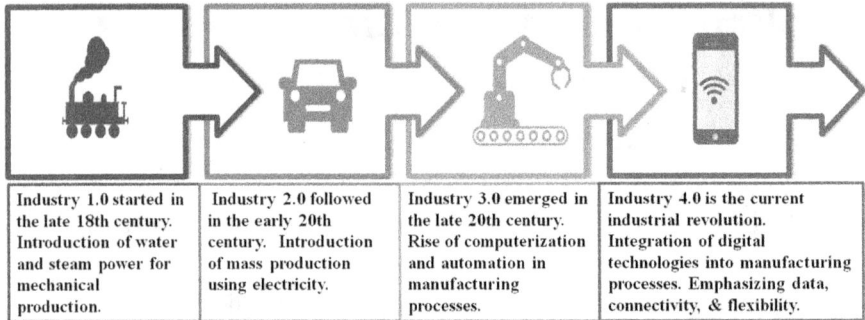

Industry 1.0 started in the late 18th century. Introduction of water and steam power for mechanical production.	Industry 2.0 followed in the early 20th century. Introduction of mass production using electricity.	Industry 3.0 emerged in the late 20th century. Rise of computerization and automation in manufacturing processes.	Industry 4.0 is the current industrial revolution. Integration of digital technologies into manufacturing processes. Emphasizing data, connectivity, & flexibility.

Figure 11.1 Schematic of industrial revolutions.

3.0 emerged in the late 20th century, with the rise of computerization and automation in manufacturing processes. Industry 4.0 is the industrial revolution that integrates digital technologies into manufacturing processes, including the Internet of Things (IoT), AI, and ML. Industry 4.0 emphasizes the importance of data, connectivity, and flexibility in manufacturing processes, leading to increased efficiency and productivity.

Human beings are continuously trying to make machines more and more autonomous and intelligent so that human effort can be reduced in every sector. Therefore, AI and ML techniques are the efforts to achieve more and more smartness among machines and to reduce human efforts. An application of these things in industries is a new generation of industry called Industry 4.0. It includes digital twin technology, Industrial Internet of Things (IIoT), 3D printing, virtual and augmented reality, big data analytics, ML, cloud computing, robots, and cobots. The new generation of telecommunication boosts the speed of these technologies, as the world is adopting the 5G network. ML algorithms make the system more and more intelligent, reliable, efficient, and flexible.

11.2 ARTIFICIAL INTELLIGENCE AND INDUSTRY 4.0

AI is one of the greatest technological achievements invented in recent years, changing human life a lot. It is a replica of human intelligence where machines can see, analyze, compute, and logically think through their data-driven experiences. Computers can carry out cognitive tasks like those performed by human minds, such as perception, thinking, learning, and problem-solving (Mckinsey, 2018). AI is a vast and multidisciplinary field of study that aims to enable machines, software, and computer-controlled robots to perform tasks that usually require human intelligence and logical reasoning. The field of AI draws upon various subjects, including computer science, biology, psychology, linguistics, and engineering.

The research started in 1956 with the invention of a perceptron (Neuron), which is similar to the human nerve system called a neural network (NN) or artificial neural network (ANN). It uses weighted sum input to simulate a human nerve system (L. Wang, 2019). AI developed the pattern of learning like a human being from big data to perform human-like cognitive things to achieve AI. Today's development in the ANN leads to solving the simplest logic to computationally very complex problems through it. It is the base of AI systems.

ML is the most widely used approach for achieving AI, and deep learning (DL) is a specific type of ML. ML and DL advancements are transforming various fields, including industrial, engineering, and technology, ushering in a paradigm shift approaching these sectors. AI is a broad field of study that aims to create intelligent machines, while ML is a part/subset of AI that focuses on teaching machines to learn from data. DL can be a specific kind of ML that use ANN to learn from data, and it is beneficial for complex tasks such as speech and image recognition. Figure 11.2 describe the relation between AI, ML, and DL.

Industry 4.0 is the great confluence of technology and manufacturing, which landed the world into a new era where data-based analytics are used for smart manufacturing. In smart manufacturing, following things are

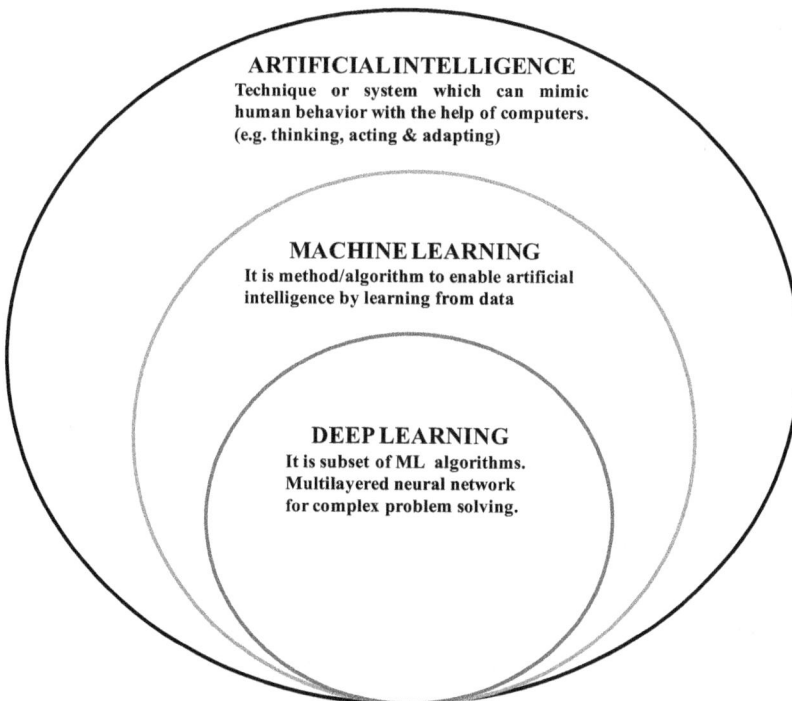

ARTIFICIAL INTELLIGENCE
Technique or system which can mimic
human behavior with the help of computers.
(e.g. thinking, acting & adapting)

MACHINE LEARNING
It is method/algorithm to enable artificial
intelligence by learning from data

DEEP LEARNING
It is subset of ML algorithms.
Multilayered neural network
for complex problem solving.

Figure 11.2 Relation between AI, ML, and DL.

included such as IIOT, digital twin, predictive maintenance, scalability, open connectivity, mass customization, edge architecture, web-based human-machine interaction (HMI), and diagnostics. It is a technology-based approach that uses internet connectivity and IoT to connect the machines to collect real-time data. This data is further processed for process optimization and condition monitoring. The aim is to completely automate the manufacturing processes to get higher efficiency, sustainability, and improved supply chain management and identify the system problems/defects before their occurrence, optimizing the process, avoiding accidents and breakdown with better safety by implementing huge volumes of data. This industrial data is processed with the help of advanced analytics to get insight into optimizing the manufacturing operations and productivity of individual assets. This advanced analytics is the subset of ML and AI.

A cyber-physical system is one of the attributes of industry 4.0 as it converges the physical, operational system, and information technology. This creates various digital solutions and advanced technologies such as digital twin, ML, IIOT, cloud computing, big data analytics and additive manufacturing (AM), augmented reality and virtual reality, and robotics. Figure 11.3 depicts the various components of Industry 4.0.

Industry 4.0 opens the doors for new opportunities as there is advancement in ML and big data analysis. It is a paradigm shift in the manufacturing sector, so the traditional manufacturing system has been replaced by smart manufacturing. A digitally interconnected physical device network is the IoT Internet called IIoT. It transmits the data and communicates with machines through the internet. IIoT system uses different sensors, software, and electronics incorporated with a machine to capture real-time online data.

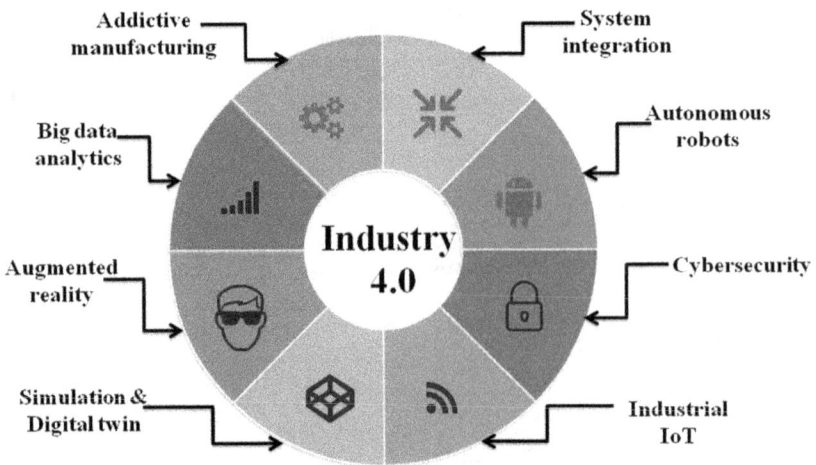

Figure 11.3 Components of Industry 4.0.

The IoT and many smart sensors are the primary enablers of curetting and storing many settings rich in industrial data relevant to all facets of production. The digital twin performs an online simulation based on information obtained from the IIoT. Digital twin illustrates a physical good, machine, procedure, or system that enables businesses to simulate their operations in real time and analyze and improve them.

The data generated by IIoT need to store somewhere. For that purpose, cloud computing is used where data can be stored, accessed, and processed with the help of an internet connection. A key enabler of the digital twin concept is cloud computing. Furthermore, cutting-edge technology like advanced robots, virtual and augmented reality, and AM enhances current-generation production. Predictive maintenance, task scheduling, process optimization, sustainability, supply chain, quality improvement, etc., are just a few of the manufacturing and production applications where ML is a subset of AI and has the potential to play a major role in revealing fine-grained complicated production patterns in the context of the smart manufacturing paradigm.

Mindful use of AI and ML technology is key to growing and driving businesses in the era of industry 4.0. Many countries worldwide are already taking advantage of these technologies in manufacturing and investing enormous amounts of money in developing them further. According to McKinsey, ML generates between \$3.5 trillion and \$5.8 trillion in value annually, or 40% of the total value that analytics today has the potential to create. Automating learning from data, identifying underlying trends, and making defensible conclusions are all possible using data-driven methodologies. Three areas of business – operations, production, and post-production – have seen success in applying ML by industries (Rai et al., 2021).

11.3 INDUSTRY 4.0 AND THE USE OF MACHINE LEARNING

Industry 4.0 integrates manufacturing systems with communicating networks and information technology to make the system intelligent. It is claimed that Industry 4.0 represents a new paradigm for intelligent and autonomous production (Jeschke et al., 2017; Y. Wang et al., 2017). Among the long list of advantages, Industry 4.0 may give manufacturing companies lucrative business plans, increased productivity, greater quality, and better working environments (Hofmann & Rüsch, 2017). Researchers and practitioners have given these possible advantages much attention (Liao et al., 2017). Physical and digital technology can be categorized under Industry 4.0. Physical technologies often refer to equipment used in production, such as sensors and drones and 3D printing technology (Gibson et al., 2015; Schwab, 2018). Most of the time, "digital technologies" refer to information and communication technologies like big data analytics, cloud computing, and simulation (Liao et al., 2017). The various industry 4.0 technologies are listed and defined in Table 11.1 (Dalenogare et al., 2018; Liao et al., 2017; Posada et al., 2015; Wan, 2015).

ML is a subset or main pillar of AI and Industry 4.0. ML can be used to predict, control, and simulate various processes. Since the past two decades, its use and research are tremendously increased due to the handling and processing of a large amount of data. Various ML algorithms can solve complex problems with the help of statistical tools and high computing power. ML is used intensively in various manufacturing areas like optimization, troubleshooting, and control (Baştanlar & Ozuysal, 2014; Pham & Afify, 2005).

Table 11.1 Different terminologies used in Industry 4.0 technologies (Bai et al., 2020).

S N	Technologies	Definition
1	Additive manufacturing (3D printing)	It is a method for building three-dimensional things by layering different materials on top of one another.
2	AI	It is a field of computer science focused on creating intelligent machines that can perform tasks typically identical to human intelligence, such as visual perception, speech recognition, decision-making, and language translation.
3	Augmented reality	Augmented reality is an interactive technology that overlays digital content on top of the real-world environment.
4	Autonomous robots (Robotics)	Autonomous robots, also known as robotics, can perform tasks autonomously without human intervention.
5	Big data and analytics	Big data and analytics refer to extracting insights and patterns from large datasets using various statistical and computational techniques.
6	Cobotic systems	Cobots, or collaborative robots, are designed to work alongside humans in a shared workspace to enhance productivity and efficiency.
7	Blockchain	Blockchain is a decentralized digital ledger technology that uses cryptography to ensure the security, transparency, and immutability of transactions and data.
8	Cloud computing	Cloud computing is a model of providing computing resources, including storage, servers, and applications, over the internet on a pay-per-use basis.
9	Cyber security	Cyber security protects computer systems, networks, and sensitive information from digital attacks, theft, and damage.

(Continued)

Table 11.1   (Continued)

S N	Technologies	Definition
10	Unmanned aerial vehicle (Drones)	Unmanned aerial vehicles, or drones, are unmanned aircraft controlled remotely or autonomously for various purposes, including surveillance, delivery, and aerial photography.
11	Global positioning system (GPS)	GPS is a worldwide satellite-based navigation system that offers highly accurate location and time data.
12	Industrial Internet of Things (IIoT)	IIoT is the application of IoT technology in industrial settings to improve operational efficiency, productivity, and safety.
13	Mobile technology	Refers to using portable electronic devices, such as smartphones and tablets, to wirelessly access and transmit information and services.
14	Nanotechnology	Nanotechnology involves manipulating materials at the nanoscale to create new materials, devices, and applications.
15	RFID	RFID, or radio frequency identification, is a technology that uses radio waves to track and identify objects in real time.
16	Sensors and actuators	These electronic devices detect and respond to physical stimuli like temperature, pressure, and motion to enable automation and data collection.
17	Simulation	Simulation is a computer-based technique to model and mimic real-world scenarios and systems to analyze and predict behavior and outcomes.

11.4 MACHINE LEARNING AND MANUFACTURING APPLICATIONS

As a part of AI, ML can learn and adjust to modifications; it presents the challenge of a rapidly changing, dynamic production line. As a result, "the systems analyst need not predict and provide options for all possible things" (Baştanlar & Ozuysal, 2014; Wuest et al., 2016). Therefore, ML offers a compelling case for why its use in manufacturing may be advantageous, given how difficult it is for most first-principle models to handle the flexibility. One of ML's key advantages is its ability to automatically learn from and adapt to changing settings (Lu, 1990). Figure 11.4 shows the general system arrangement of the basic ML setup.

11.5 ML ALGORITHMS

ML performs the task with the help of various algorithms. These algorithms are the scientific, logical steps to solve the problem or find the pattern in the dataset, as basic steps shown in Figure 11.4. Therefore, ML can be defined

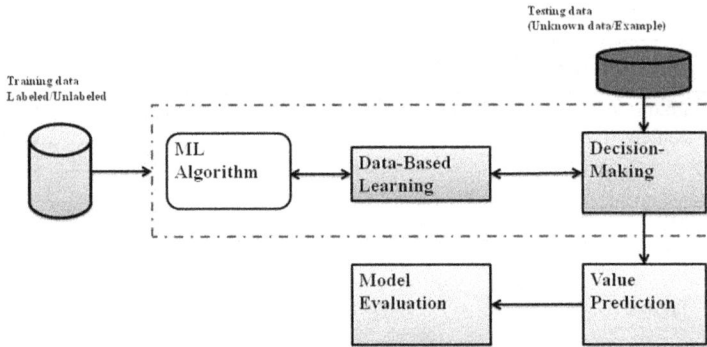

Figure 11.4 Basic ML system arrangement (Nti et al., 2022).

as a scientific inquiry into the use of algorithms or computational models in computers, which make accurate predictions and use experiences to improve performance on a job (Wuest et al., 2016; Priore et al., 2001). Experience can be the chronological data used to forecast and create the model.

ML algorithms can be classified as supervised learning (SL), unsupervised learning (UL), and reinforcement learning (RL). SL algorithms use labeled data where the input and output of the data are known. In SL, users provide known dataset along with its output to train the mode. This model is used for solving a similar type of dataset but of unknown output, where the algorithm knows the previous example, and on that, it can solve for this similar dataset.

Figure 11.5 describes the various types of ML techniques and categories of ML with their learning tasks, such as SL used for classification and regression and UL for clustering and association rule. Similarly, reinforcement learning is of two ways, either model-free or model-based.

UL is the second type of ML learning algorithm where data is unlabeled. This means the dataset is completely unknown. In this dataset, algorithms try to find the pattern or specific cluster in the given dataset.

Reinforcement learning (RL), where the environment plays an important role in which algorithms interact with the environment and logically infer the features from the dataset without supervisory learning, where interaction with the environment is the only teacher. This algorithm is used for robotic navigation, real-time decision-making, and the learning of machine skills (Sharma et al., 2020). Figure 11.5 shows the types of ML algorithms and their learning task. ML can help the manufacturing sector identify patterns in current datasets that can be used as the foundation for predicting how well a system will perform in the future (Baştanlar & Ozuysal, 2014; Nilsson, 2005). New knowledge acquired through ML can assist process managers in making informed judgments or be automatically integrated into the system to enhance its performance. Many ML methods aim to identify distinctive regularities or patterns that define interactions (Baştanlar & Ozuysal, 2014).

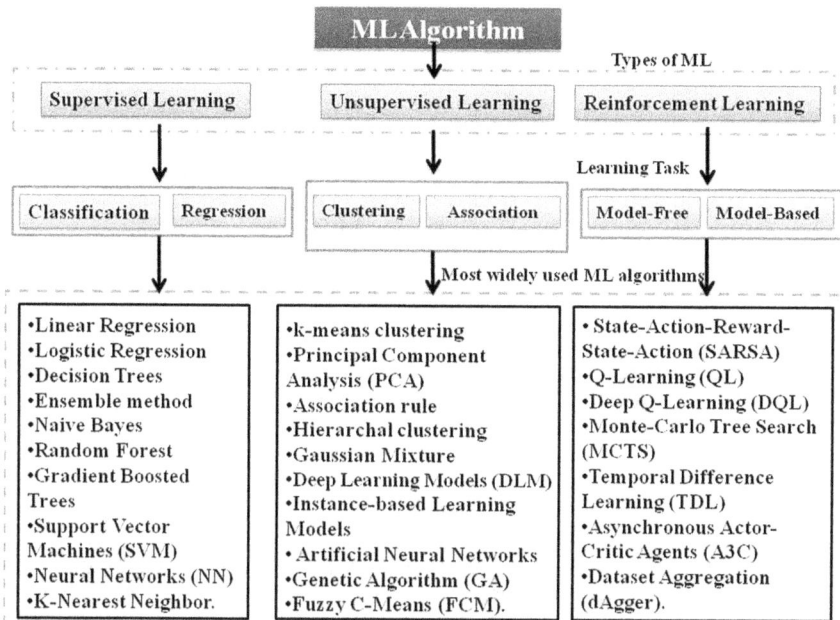

Figure 11.5 Types of ML algorithms (Nti et al., 2022).

The main purpose of ML algorithms is to churn out knowledge from a given dataset (Kwak & Kim, 2012). Therefore, it is important to recognize that collected data becomes valuable. For that, data should be analyzed and transformed into actionable knowledge, which can be utilized, such as to generate predictions (Learning, 2010). This is particularly true for the manufacturing industries, where limitations in understanding, funding, and technology make it challenging to gather real-time datasets during the execution of a live production system. This might impact the decision of where to put process milestones (Wuest et al., 2016). While carefully choosing checkpoints makes sense from the standpoint of what data points are helpful, it can be superfluous, given the logical capacity of ML approaches to mine information from previously thought of using less dataset. This could make it possible to identify more states and collect data throughout the manufacturing operation. It has to be investigated if this is advantageous or not. The technical aspect of assessing the new data presents no issue, given that ML can handle high-dimensional data. The capacity to acquire the data, especially, may still be a difficulty in data acquisition. Finding state drivers in situations with very high dimensionality isn't anticipated to be difficult once the data are available. Table 11.2 gives an overlook of the ability of ML to give solutions to various manufacturing applications.

ML algorithms have various applications in manufacturing and machining optimization processes. The selection of the appropriate ML algorithm

Table 11.2 Feasibility of ML models in the manufacturing sector.

Requirements of the Manufacturing Sector	Machine learning's Theoretical Capability to Meet Requirements
The capacity to reasonably manage high-dimensional problems and datasets.	Certain ML methods, such as SVM can handle high dimensionality (>1000) effectively, but concerns of probable overfitting must be measured (Widodo & Yang, 2007).
Ability to simplify results and offer practitioners transparent guidance that they can act on.	ML can automatically enhance a system, create approximations of future behavior, and extract patterns from existing data (Kubat, 2017).
The capacity to change with environments while exerting acceptable effort and money.	The ability of ML, a component of AI, to learn and adjust to changes removes the need for system designers to provide solutions for every scenario that might arise (Kubat, 2017).
Ability to advance existing knowledge by learning from results.	ML can create new information and knowledge by identifying patterns in existing data (Pham & Afify, 2005).
Working with manufacturing data already accessible without requiring to capture specific data right away.	ML techniques aim to extract information from already-existing data (Kwak & Kim, 2012), and that stored data is only useful after analysis and transformation into knowledge that can be used for predictions (Kubat, 2017).
Understanding of relevant process inter- and intra-relationships, preferably with correlation or causality.	Some ML methods look for relations-related patterns or regularities (Kubat, 2017).

for a specific application depends on several factors, such as the volume and type of data, the complexity of the relationships between variables, and the level of accuracy required. Table 11.3 shows the various ML algorithms, their applications in the manufacturing sector, and their strengths and weaknesses. Table 11.3 gives an overall idea of various ML algorithms and their potential in manufacturing.

Table 11.3 shows that the SVM algorithm is a popular choice for manufacturing and machining optimization processes due to its memory-efficient algorithm and effectiveness in high-dimensional spaces. The decision tree algorithm is also widely used due to its simplicity, ease of implementation, and model interpretability. Regression analysis is commonly used for forecasting and optimization problems, while ANN is suitable for larger volumes of data and more complex relationships. However, ANN's development time and computational cost are higher than other algorithms. Clustering algorithms are effective in UL methods, while Bayesian networks are useful for determining causal relationships between random variables. Ensemble learning algorithms improve ML results by combining multiple models, but they are difficult to interpret. DL algorithms are suitable for more complex

Table 11.3 Machine learning algorithms for various manufacturing applications.

ML Algorithm	Description	Strengths	Weakness	Application Area in Manufacturing
Decision Tree (Baştanlar & Ozuysal, 2014)	It is a classification algorithm that classifies larger datasets into smaller datasets.	a) Scaling and normalization of data is not necessary. b) Involves minimal data preparation and pre-processing efforts. c) Easy to explain to stakeholders.	a) Minor data changes impact decision tree structure greatly. b) Inadequate for continuous value prediction and regression. c) More complex calculations than other algorithms.	Yield enhancement, technology selection, life cycle engineering, quality management, green design.
Regression analysis (Kubat, 2017)	RA is a predictive model that uses equations to represent the relation between input and output parameters.	a) Regularization is needed to avoid overfitting. b) Implementation is straightforward.	a) May underfit and unsuitable for complex relationships. b) Outliers can significantly impact the results.	Machining parametric optimization, forecasting, performance measurement and evaluation.
Support vector machine (SVM) (Baştanlar & Ozuysal, 2014; Kubat, 2017)	A boundary detection algorithm that identifies different data points in multidimensional boundaries.	a) Advantageous in situations where the data has a high number of dimensions and a relatively small number of samples. b) Memory-efficient algorithm.	a) Not well-suited for larger datasets or data with high levels of noise. b) Less effective for high-dimensional data.	Quality assessment, process planning, tool wear prediction, pattern recognition, and forecasting.

(Continued)

Table 11.3 (Continued)

ML Algorithm	Description	Strengths	Weakness	Application Area in Manufacturing
Artificial Neural Networks (ANN) (Kubat, 2017; Learning, 2010)	A computational and mathematical model that draws inspiration from nervous systems.	a) No restriction on how the input variables are distributed. b) Can be applied to complex relationships and a big volume of data.	a) Requires more data than other algorithms. b) Longer model development time. c) Computationally costly.	Capacity utilization, demand forecasting, quality improvement, machining parametric optimization.
Clustering (Baştanlar & Ozuysal, 2014; Kubat, 2017)	Divides data into k clusters, where k is determined by the clustering algorithm.	a) Straightforward implementation without distributional assumptions. b) Efficient unsupervised learning with larger labeled datasets.	a) Ineffective for intricate geometric shapes. b) Disregards distant data points within the same cluster.	Engineering design, manufacturing system design, quality assurance, production and process planning.
Bayesian Networks (Kubat, 2017; Learning, 2010)	Predicts the result class using Bayes' theorem. Prior probability and conditional probability are both utilized.	a) Suitable for establishing causal links between random factors. b) A single analysis can include both direct and indirect evidence.	a) Models may be complex. b) Computationally costly. c) Ineffective for high-dimensional data.	Performance evaluation, monitoring and diagnosis, fault diagnosis, resilience modeling.

Technique	Description	Advantages	Disadvantages	Applications
Genetic Algorithms (Kubat, 2017; Learning, 2010)	Genetic algorithms apply natural selection as a search heuristic to optimize various problems. They evolve potential solutions via iterations and breed the fittest for further selection.	a) Robust and adaptable, excels at multi-objective optimization and noisy environments. b) Finds solutions from a population of points, not a single one.	a) Computationally costly and time-intensive. b) Demands less information, posing challenges in objective function design.	Parameter optimization, planning and scheduling, supply chain management, etc.
Instance-based learning (Kubat, 2017)	This algorithm compares new data with past instances stored in memory to make predictions or classifications.	a) Concentrates on the objective function's local approximation. b) Easily adaptable to new data alterations.	a) High computational cost. b) Demands substantial memory for data storage.	Simulation and animations, flexible manufacturing, system design.
Ensemble Learning (Kubat, 2017)	Solves computational intelligence difficulties by using several models to improve the forecast, classification of data and function approximation.	a) It enhances ML outcomes by combining various models to produce a more precise forecast model. b) Less noise levels.	a) High computational cost. b) Difficult to interpret.	System designing, MRR predictions, supplier selection, and system diagnosis.
DL (Kubat, 2017; Learning, 2010)	DL is ML subset that uses artificial neural networks (ANNs), categorized into three types: deep belief networks, deep recurrent NNs, and stacked auto encoders.	a) Adaptable to system changes. b) Enables parallel data processing for faster computations. c) Applicable across various data types.	a) Requires significant amounts of dataset for accuracy. b) High computational cost. c) Demands higher skill levels.	Quality analysis, fault diagnosis, and distortion prediction in 3D printing.

problems but require larger volumes of data and higher skill levels. Genetic algorithms are effective in optimization problems but are computationally expensive.

In summary, the appropriate selection of the ML algorithm for a specific application depends on various factors, and different ML algorithms have their strengths and weaknesses. Therefore, careful consideration should be given to selecting the appropriate ML algorithm for a specific manufacturing and machining optimization process.

11.6 ML FRAMEWORK FOR MANUFACTURING

Since one of ML's primary objectives is to enable the more intelligent use of goods and services, optimize processes in the industrial sector by enabling more intelligent use of goods and services. ML can reduce wastage while reducing costs and saving time in the industrial sector. It also makes it possible to create algorithms for controlling human behavior. ML is used in various manufacturing problems, from conventional to non-conventional. As discussed earlier, data acquisition to perform the ML techniques is critical. Every ML technique works on data, and that data should be proper and preprocessed to get accurate results. Therefore, when considering the complete framework for using ML in the manufacturing sector, it needs to start with data collection. For data collection, various sensors and data acquisition systems are used.

To implement ML algorithms in the manufacturing problem-solving process, the steps shown in Figure 11.6 can be followed. Several iterations are usually required in the ML analysis process to achieve better outcomes. It begins with clearly defining the issue and selecting an appropriate analysis technique. Next, necessary data is collected and preprocessed to enable analysis. An ML model is created, selected, and assessed, and findings are examined to develop a solution for the issue (Cho & Kang, 2016).

In manufacturing applications, a framework refers to a pre-designed structure or set of tools that provide a foundation for building and deploying ML models in a manufacturing setting. An ML framework typically includes libraries and tools for data preparation, model training, model evaluation, and deployment. It can also include pre-built models or modules that can be customized for specific manufacturing applications. These frameworks provide a high-level interface for building and deploying ML models, making it easier for engineers and data scientists to develop ML

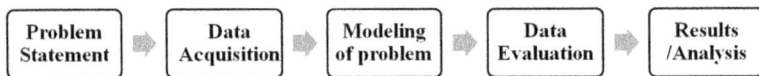

Problem Statement ➡ Data Acquisition ➡ Modeling of problem ➡ Data Evaluation ➡ Results /Analysis

Figure 11.6 General flow of problem-solving using ML implementation.

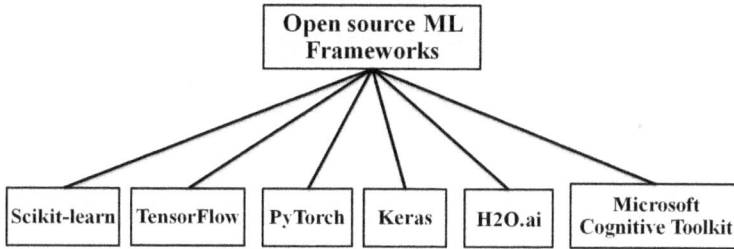

Figure 11.7 Open-source machine learning frameworks.

solutions for manufacturing challenges such as quality control, predictive maintenance, and process optimization. An ML framework can also improve model accuracy, faster development, and better scalability.

Some examples of popular ML frameworks for manufacturing applications include TensorFlow, Keras, and PyTorch. Figure 11.7 displays open-source free frameworks for ML that can be used for manufacturing applications (Khan & Al-Badi, 2020). These frameworks are available for designing, developing, training, and testing ML models in industrial environments. The choice of ML framework is crucial for the success of the project. Various ML frameworks are available for developing models, each with strengths and limitations. It is essential to consider these factors when selecting the appropriate framework for a particular application.

Several ML frameworks can be used for manufacturing applications. Here are a few popular ones:

- TensorFlow: TensorFlow is a popular open-source ML framework developed by Google Inc. USA. It is widely used in manufacturing for predictive maintenance, quality control, and anomaly detection tasks.
- PyTorch: PyTorch is another popular open-source ML framework widely used in manufacturing. It is known for its ease of use and flexibility and is often used for tasks such as computer vision and natural language processing.
- Keras: Keras is a high-level ML framework built on top of TensorFlow. It is known for its ease of use and allows developers to build and deploy ML models quickly.
- Scikit-learn: Scikit-learn is a popular ML library built on NumPy and SciPy. It is widely used for classification, regression, and clustering tasks in manufacturing applications.
- H2O.ai: H2O.ai is an open-source ML platform designed to be easy to use and scalable. It is often used in manufacturing for predictive maintenance and quality control tasks.
- Microsoft Cognitive Toolkit: Microsoft owns the open-source DL programming environment known as the Cognitive Toolkit. It works with recurrent neural networks (RNN), feed-forward deep neural networks

(DNN), and convolution neural networks (CNN) and enables multi-machine, multi-GPU back-ends. It was initially developed to create DL models by simulating the anatomy of the human brain. Companies and groups can use this toolkit to investigate ML options, such as aircraft predictive maintenance (Khan & Al-Badi, 2020). When choosing an ML framework for manufacturing, it is important to consider factors such as ease of use, scalability, and compatibility with existing systems.

Implementing ML frameworks in industrial settings involves generating data through sensors, different actuators, and mobile applications in manufacturing industries. The generated data is then stored either locally or in the cloud for further processing. Once the data is cleaned, different ML models such as TensorFlow and Torch are applied to the datasets using support vector machine, linear regression and decision tree algorithms. The model that generates the highest score in the confusion matrix is chosen to predict new datasets. Decision-makers then utilize the output from the ML model to inform their decision-making processes. Figure 11.8 shows the general framework for implementing ML in the manufacturing industry for better decision-making.

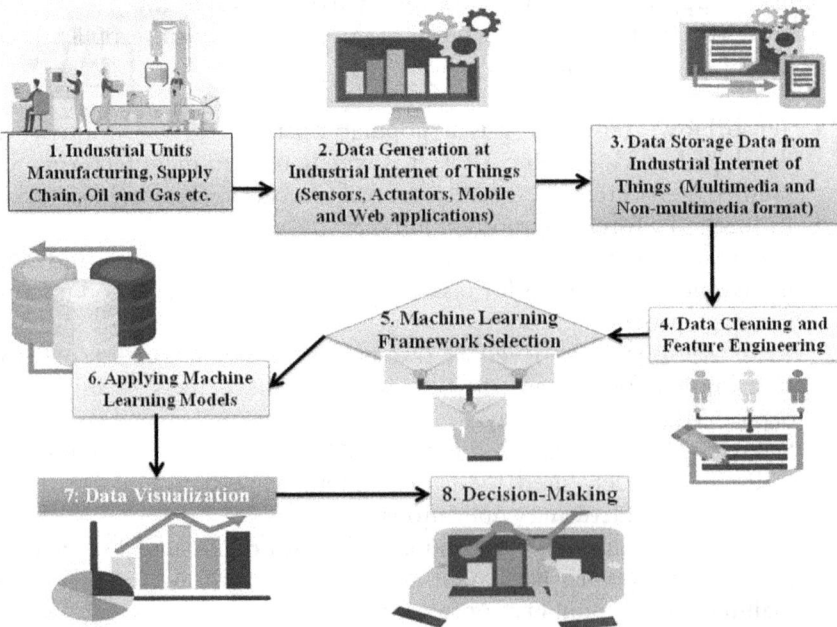

Figure 11.8 Framework for implementation of ML in an industrial setting (Khan & Al-Badi, 2020).

11.6.1 Machining processes using machine learning

The current Fourth Industrial Revolution is the union of the digital and physical worlds, opening the door for innovations in many areas, including AI, robotics, 3D printing, autonomous cars, quantum computing, nanotechnology, and the IoT. ML has changed the course of AI during this era. ML has brought a new age of "smart machining," enabling machines to learn from data and improve without explicit programming. Smart machining involves integrating machine tools through a cyber-physical system to improve product quality, increase output, track system health, and optimize design and process parameters (Kim et al., 2018). This section summarizes machining procedures that utilize ML algorithms to present a standpoint on the machining process industry.

To accomplish specific objectives, smart machining/cutting is an advanced manufacturing process that can automatically modify its parameters. Unlike traditional machining methods, smart machining can overcome errors during material removal, such as dimensional errors, elastic and thermal deformations, and vibration issues. By integrating various systems like machine tools, sensors, big data, cloud-based systems, simulation-based designs, and smart control algorithms, smart machining can enhance processing efficiency and product quality through real-time process parameter optimization. Workpiece properties, machine parameters, cutting tools, and cutting conditions are all critical factors affecting the final product's quality. Additionally, since handling and positioning operations take up more than half of the total working time, it's critical to optimize control parameters during these operations (Mekid et al., 2009).

Several researchers covered the application of ML in a variety of manufacturing industries. Table 11.4 provides examples of machining processes that use ML techniques. While the working principles of various ML algorithms are well known, this section only summarizes the specifics of their implementation in the context of machining processes.

ML has been utilized to improve traditional and non-traditional machining techniques for prognostics, product quality forecasting, diagnostics, and parameter optimization resulting in cost-effective production for the manufacturing industry. The most commonly used algorithms were SVM and ANN, which performed well (Kim et al., 2018). However, the preprocessed data and extracted features serve as input parameters for further analysis and modeling, significantly impacting these algorithms' accuracy. To improve the practical implementation of these algorithms, conducting further research specifically on feature extraction techniques will be crucial. In addition, ANN and SVR have been employed to enhance the machine's structure, and thermal and dynamic characteristics. However, there is insufficient information available regarding the performance of these algorithms, and it is necessary to compare their results with experimental data to determine which algorithm is more suitable for the task (Kim et al., 2018). Most of the

Table 11.4 Application of machine learning in smart machining processes.

(a) Conventional Machining

Purpose	Input Parameters	Algorithms	Pre-Processing	Accuracy & Ref. (Year)
Milling				
Tool breakage detection	Power consumption data and cutting force	SVM, SVR	----------	99.38% (Cho et al., 2005)
Tools wear prediction	Vibration data, cutting force, and acoustic emission	RF	Mean, standard deviation, median	99.20% (Wu et al., 2017)
Prediction of remaining useful life (RUL) and tool wear	Vibration data, acoustic emissions, and cutting force	SVR	Isometric feature mapping, wavelet packet decomposition, principal component analysis (PCA)	98.95% (Benkedjouh et al., 2015)
Monitoring of tool wear	Tool images	SVM and K-NN	Contour descriptors and image shape	90.26% (Kim et al., 2018)
Surface roughness prediction	Feed rate, depth of cut, spindle speed	SVM	Normalization of data	86.50% (X. Lu et al., 2016)
Energy consumption prediction	Depth of cut, feed rate, spindle speed, cutting strategy, tool axis	Gaussian process regression	----------	Above 95% (Park et al., 2015)
Energy consumption prediction	feed rate, spindle speed, depth of cut, tool cutting direction, length of tool path, cutting strategy	GPR (global and collective)	Gaussian mixture method	98.66% (global GPR) (Bhinge et al., 2015)
Condition monitoring of tool	Vibration signals (from accelerometer)	Feed forward BpNN, J48 decision Tree	Fast Fourier transform (FFT), mean, standard deviation, kurtosis	95.40% (NN), 94.30% (J48), (Krishnakumar et al., 2015)

Turning

Application	Input parameters	Method	Technique	Accuracy / Reference
Machining parameters prediction	Cutting speed, feed rate, and depth of cut	ANN, SVR, polynomial regression (PR)	Regularization	92.48% (PR), 93.15% (ANNs) (Jurkovic et al., 2018)
Parametric optimization and prediction of carbon emission	Feed rate, speed, depth of cut	Regression, MOTLBO	Grey relational analysis (GRA), response surface method (RSM)	Above 95% (Lin et al., 2015)
Surface roughness prediction	Feed rate, speed, depth of cut, vibration, flank wear	Multiple linear regression (MLR)	PCA and mean, median, standard error, Kurtosis, etc.	80.8% (Elangovan et al., 2015)
Prediction of grain size and micro hardness	Feed rate, cutting speed, tool coating status, tool edge radius	RF, GA	-----------	96.5% (Laha et al., 2015)
Condition monitoring of tool	Vibration signals	K-Star classifier algorithm	Correlation-based attribute subset selection, and standard deviation and error, variance, Kurtosis, etc.	78.69% (Painuli et al., 2014)
Tool life estimation	Depth of cut, speed, feed rate, temperature	BpNN, regression analysis method	-----------	N/A (Tosun & Özler, 2002)
Pattern recognition and tool wear prediction	White pixels of tool image, cutting speed, machining time, depth of cut, feed rate	DNA-based computing, cascade forward BpNN	Binary image data, and image processing	75.0% (D'Addona et al., 2017)
Online tool life prediction	Six signal characteristics from a sound, vibration, and cutting force sensor	Cascade-forward NN, feed-forward NN	Wavelet feature extraction	N/A (Karam et al., 2016)

(Continued)

Table 11.4 (Continued)

Purpose	Input Parameters	Algorithms	Pre-Processing	Accuracy & Ref. (Year)
Drilling				
Estimation of geometric profile and quality	Cutting force, thrust force, t	Logical analysis of data	------------	94.60% (Shaban et al., 2015)
Boring				
Prediction of chatter	Depth of cut, spindle speed, feed rate	SVM	Discrete wavelet transforms method	95.0% (Saravanamurugan et al., 2017)
Grinding				
Monitoring of surface shape peak valley (PV) and surface roughness (Ra)	Acoustic emission, vibration, grinding force	IFSVR	Fast Fourier transform and identification model	75.93% (PV), 85.19% (Ra), (Zhang et al., 2015)
(b) Non-Conventional Machining				
Abrasive Water Jet				
Prediction of surface roughness	Material thickness, cutting speed, measurement position, and abrasive flow	Extreme ML, ANN, GPR.	----------	96.65% (Čojbašić et al., 2016)
Prediction of surface roughness	Water jet pressure, traverse speed, abrasive flow rate, and abrasive grit size	SVM	GRA	99% (Mat Deris et al., 2013)
Surface roughness prediction	Stand-off distance, water jet pressure, traverse speed, abrasive flow rate, and abrasive grit size	Feed-Forward BpNN, Regression model	----------	96.99% (NN), 99% (regression) (Çaydaş & Hasçalik, 2008)

Electrochemical Discharge Machining (ECDM), Electrochemical Machining (ECM)

Application	Parameters	ML technique	Result (Reference)
Parametric optimization to get maximizing MRR & minimizing radial overcut	Electrolyte concentration, applied voltage, Electrolyte flow rate, inter-electrode gap	TLBO	18% improvement in MRR (Rao & Kalyankar, 2011)
Laser Machining			
Predict the output, dimensional characteristics, and surface quality of laser-cut microchannels	Scanning speed, pulse intensity, and pulse frequency	NN, decision tree, KNN, linear regression	88.7% (NN), 76.90% (decision tree) (Teixidor et al., 2015)
Electric Discharge Machining (EDM)			
Parametric optimization for maximum MRR & minimum wear ratio	Pulse on/off time, and pulse current	BpNN, particle swarm optimization, simulated annealing, GA	N/A (Majumder, 2015)
Parametric optimization for maximum MRR and MRR estimation	Gap voltage, capacitance, speed, and feed rate	Feed-forward BpNN, GA	96.06% (Somashekhar et al., 2010)
Machining parameter optimization minimum surface roughness and for maximum MRR	On time, off time, and mean current	GPR, NSGA-II	N/A (Yuan et al., 2008)
Machining parameter optimization for minimum surface roughness and maximum MRR	On/off time, cutting radius, wire feed, voltage, Arc on/off time, water flow	GRA	N/A (Chiang & Chang, 2006)

case studies reviewed in the table were published in the past ten years; it indicates that the domain of ML is actively researched. Thus, it is reasonable to anticipate an increase in machining process cases utilizing ML. Integrating ML algorithms in machining processes has enormous potential for improving manufacturing efficiency and product quality. The study highlights the importance of feature extraction techniques in enhancing the accuracy of ML algorithms. Nonetheless, more research is needed to verify the algorithms' performance and compare them with experimental data. As the field of ML continues to evolve, more case studies are expected to emerge, paving the way for innovative and cost-effective manufacturing processes.

11.7 SCOPE AND FUTURE DIRECTIONS

ML has various applications in the manufacturing industry, including quality control, predictive maintenance, supply chain management, and process optimization. As technology advances, the scope and future of ML in manufacturing are significant. One of the primary areas where ML can be applied is predictive maintenance. ML algorithms can evaluate huge amounts of data from sensors and other sources to predict when equipment is expected to fail, allowing manufacturers to schedule maintenance proactively. This can lead to reduced downtime, improved efficiency, and cost savings for manufacturers. ML can also be used for quality control by analyzing images and identifying product defects. By automating this process, manufacturers can improve product quality and reduce the risk of recalls or returns, saving time and money. Supply chain management is another area where ML can optimize processes. By analyzing supplier performance, demand, and inventory levels, manufacturers can make more informed decisions about sourcing and logistics, reducing waste and improving efficiency (Wuest et al., 2016).

ML can also permit new forms of manufacturing, such as AM (3D printing). By using ML algorithms to design and optimize parts and prototypes, manufacturers can reduce design and production time and improve the manufacturing process. The prospect of ML in the manufacturing industry is promising. As technology advances, manufacturers can collect and analyze even more data, leading to more accurate predictions and improved decision-making. ML will also become more accessible, making tools and platforms more user-friendly and affordable.

The scope and future of ML in the manufacturing industry are significant. ML has the potential to revolutionize manufacturing by improving efficiency, reducing costs, and improving product quality. Manufacturers that invest in ML today will be well-positioned to reap the benefits in the years to come. AI and ML can potentially revolutionize the manufacturing industry, and their impact will only increase as technology advances. Manufacturers must stay ahead of the curve and invest in AI and ML to remain competitive in an increasingly global market. Manufacturers must invest in infrastructure, data analytics, and talent to fully realize AI and ML's benefits. Data privacy and security will also be critical as more data is collected and analyzed. Additionally, manufacturers must navigate regulatory and ethical considerations when implementing ML.

11.7.1 Machine learning advantages for manufacturing industry

ML techniques have successfully predicted maintenance requirements across various sectors and optimized, monitored, and controlled various production processes (Kubat, 2017; Pham & Afify, 2005; Kwak & Kim, 2012; Susto et al., 2015). In particular, ML has demonstrated significant promise for enhancing quality control in manufacturing applications (Apte et al., 1993), especially in complex manufacturing scenarios where it can be difficult to pinpoint the underlying causes of issues (Harding et al., 2006). These ML applications, however, frequently have a narrow focus, concentrating on particular processes rather than the complete manufacturing system (Doltsinis et al., 2018). ML techniques are expert at handling multivariate and high-dimensional data, extracting implicit relationships from large, dynamic, and complex datasets, even in messy environments (Köksal et al., 2011). Because there is typically more data than expertise in engineering and manufacturing problems, ML provides a way to improve understanding in this field (S. C. Y. Lu, 1990). While the general benefits of ML are highlighted in this section, it's essential to remember that the particular benefits can change depending on the ML technique used. ML is generally well known for its capacity to cut cycle time and scrap and increase resource utilization for challenging manufacturing issues. Moreover, ML offers valuable tools for enhancing continuous quality in challenging and extensive processes (Monostori, 1993; Pham & Afify, 2005).

ML algorithms offer an improvement in handling high-dimensional problems and data, which is particularly valuable, given the increasing accessibility of intricate data (Yu & Liu, 2003) and the lack of transparency in manufacturing processes (Wuest et al., 2016). However, it's important to note that this advantage cannot be universal since several algorithms, such as SVM and distributed hierarchical decision tree, are better suited for high-dimensional data than others (Bar-Or et al., 2005; Do et al., 2010). ML algorithms' applicability in manufacturing mainly depends on their ability to manage high-dimensional data. Consequently, the potential to handle high dimensionality is a benefit of utilizing ML in manufacturing.

The success of ML algorithms in achieving results within manufacturing environments has been well-documented, although the data requirements may vary depending on the type of algorithm used, such as whether it is supervised, unsupervised, or involves reinforcement learning (Filipič & Junkar, 2000; Guo et al., 2008; D. Kim et al., 2012; Nilsson, 2005).

Manufacturing systems are characterized by their complexity, uncertainty, and dynamic nature to some degree. ML algorithms give the prospect to automatically learn from these dynamic systems and adapt to altering environments (Lu, 1990; Simon, 1983). Therefore, the speed of the adaptation process can vary depending on the particular ML algorithm used, but it is typically quicker than more conventional techniques. The function of ML in manufacturing can result from patterns in already-existing datasets, establishing a foundation for approximations regarding the system's future behavior (Nilsson, 2005). This new information can either assist process decision-makers in their decision-making or enhance the system automatically. Some ML methods are designed to find relationships in the data by looking for patterns or regularities (Baştanlar & Ozuysal, 2014). The efficacy of various ML algorithms in manufacturing applications was compared by scientists (Konieczny & Idczak, 2016; Wuest et al., 2016). However, it should be noted that this is only a rough guide and not a set of stringent rules for choosing an appropriate ML algorithm.

11.7.2 Machine learning implementation challenges in manufacturing industry

In manufacturing, ML applications face a sizable task in acquiring relevant data. Since the quality, quantity, and structure of the production data significantly affect the performance of ML algorithms, this issue also serves as a constraint. The data collection may offer several difficulties, such as high-dimensional data that may contain redundant and unnecessary information and hamper the system performance of the learning algorithms (Yu & Liu, 2003).

Many ML methods currently utilized are designed to operate solely with data with fixed and arbitrary values (Pham & Afify, 2005). Numerous variables, such as the algorithm and parameter values, affect how much data affects ML algorithms. Due to security issues or lack of data capture during the operation, obtaining data is generally a struggle for most manufacturing studies. Despite ML's capacity to derive knowledge and produce superior results with fewer data than conventional techniques, some data-related factors may prevent its effective implementation. Understanding the data is, therefore, essential for AI applications. The author stressed that while conventional methods concentrate on information extraction, ML requires substantial time and work to prepare the data (Wuest et al., 2016).

Pre-processing greatly impacts the outcomes, which is frequently necessary once the accessible data is secured for ML application. Data normalization and filtration are two typical pre-processing tasks that standardized tools can support. Checking for irregular training data that can affect the training of particular algorithms is also essential. Missing or inaccessible data points in the data collection are a frequent issue in manufacturing practice. The implementation of ML systems is hampered by these "missing values" (Pham & Afify, 2005). Choosing the appropriate ML technique and algorithm is a major challenge becoming more essential. The range of troubles and their different requirements emphasizes the necessity of specialized algorithms that possess unique strengths and limitations (Wuest et al., 2016). There are now many ML algorithms and their variations available due to increased interest in the domain of ML application in manufacturing from practitioners and academics. In addition, "hybrid approaches", which combine multiple algorithms, are spreading, increasing the complexity already present, and offering superior outcomes to using a single algorithm alone (Lee & Ha, 2009).

Despite numerous studies demonstrating the effective use of ML methods for particular problems, the absence of publicly accessible test data frequently makes it difficult to have an unbiased and neutral assessment of the findings, impeding a final evaluation. The currently accepted procedure includes the following steps to choose appropriate ML algorithms for a specific problem. To select an appropriate ML method, the first step is to analyze the available data and its description to determine whether a supervised, unsupervised, or RL method should be used based on the availability of labeled, unlabeled, or expert knowledge. Next, evaluate the common applicability of existing algorithms for the research problem, including their ability to handle high dimensionality (Do et al., 2010), structure, data types, and quantity of data available for training and evaluation. Finally, to identify a suitable algorithm, explore previous applications of algorithms on similar problems in other domains (Wuest et al., 2016).

Currently, supervised algorithms are prevalent in manufacturing due to the availability of labeled data. However, increasing data availability may lead to unsupervised methods, including RL, gaining more significance in the future. Due to advancements in big data, hybrid approaches that combine the advantages of various ML techniques have been launched and are receiving more attention. The importance of ML, an efficient tool for intelligent and smart manufacturing, is anticipated to grow in the coming years. However, progress in the field depends on interdisciplinary cooperation between various fields like computer science, statistics, industrial engineering, mathematics, and mechanical and electrical engineering (Widodo & Yang, 2007; Wuest et al., 2016).

11.8 CONCLUSIONS

The growth of ML in the manufacturing sector has been facilitated by the rapid development of algorithms, increased data availability, and enhanced computing power. The implementation of ML for manufacturing has seen a significant increase in current years due to the dawn of Industry 4.0 and the widespread use of AI. This chapter has provided a comprehensive overview of the application of ML and AI in the manufacturing sector. The chapter started with an introduction to the concept of ML and its significance in the modern era of Industry 4.0. The rise of AI has been crucial in adopting ML in the manufacturing sector. Industry 4.0, which emphasizes the digitalization of manufacturing processes, has provided an ideal environment for ML to thrive. The chapter explored the various applications of ML in manufacturing. ML has effectively solved manufacturing-related issues by analyzing vast amounts of data generated by sensors, actuators, and mobile applications. The use of ML in decision-making processes has also been highlighted, as it helps decision-makers to make informed choices.

The chapter provided a framework for implementing ML models for the manufacturing sector, including selecting appropriate algorithms, such as linear regression, support vector machine (SVM), and decision tree, and the applications of ML models, such as TensorFlow and Torch. The framework ensures that the ML models produce the best results, which are then utilized to inform decision-making processes.

A use case and discussion were presented in the chapter, which demonstrated the effectiveness of ML in the machining processes. The chapter provided insights into ML's future direction and scope in manufacturing. As the manufacturing sector continues to evolve, ML will play a more and more significant role in driving innovation and improving efficiency. The benefits of the widespread adoption of ML in the manufacturing industry include cost reduction, increased productivity, and better overall performance.

As the manufacturing sector continues to evolve and embrace digital transformation, ML will undoubtedly play a crucial role in driving innovation and improving efficiency. In conclusion, this chapter has demonstrated that ML is a powerful tool for solving manufacturing issues and improving efficiency. Its applications in the manufacturing sector are numerous and far-reaching, and its potential for future development is limitless. The widespread adoption of ML in the manufacturing sector will drive innovation and lead to significant benefits for manufacturers, consumers, and society as a whole.

REFERENCES

Apte, C., Weiss, S., & Grout, G. (1993, June). Predicting defects in disk drive manufacturing. A case study in high-dimensional classification. *Proceedings of the Conference on Artificial Intelligence Applications*, 212–218. https://doi.org/10.1109/caia.1993.366608

Bai, C., Dallasega, P., Orzes, G., & Sarkis, J. (2020). Industry 4.0 technologies assessment: A sustainability perspective. *International Journal of Production Economics*, 229, 107776. https://doi.org/10.1016/j.ijpe.2020.107776

Bar-Or, A., Schuster, A., Wolff, R., & Keren, D. (2005). Decision tree induction in high dimensional, hierarchically distributed databases. *Proceedings of the 2005 SIAM International Conference on Data Mining, SDM 2005*, 466–470. https://doi.org/10.1137/1.9781611972757.42

Baştanlar, Y., & Ozuysal, M. (2014). Introduction to machine learning, second edition. In *Methods in Molecular Biology (Clifton, N.J.)* (Vol. 1107). https://doi.org/10.1007/978-1-62703-748-8_7

Benkedjouh, T., Medjaher, K., Zerhouni, N., & Rechak, S. (2015). Health assessment and life prediction of cutting tools based on support vector regression. *Journal of Intelligent Manufacturing, 26*(2), 213–223. https://doi.org/10.1007/s10845-013-0774-6

Bhinge, R., Biswas, N., Dornfeld, D., Park, J., Law, K. H., Helu, M., & Rachuri, S. (2015). An intelligent machine monitoring system for energy prediction using a Gaussian Process regression. *Proceedings - 2014 IEEE International Conference on Big Data, IEEE Big Data 2014* (pp. 978–986). https://doi.org/10.1109/BigData.2014.7004331

Çaydaş, U., & Hasçalik, A. (2008). A study on surface roughness in abrasive water-jet machining process using artificial neural networks and regression analysis method. *Journal of Materials Processing Technology, 202*(1–3), 574–582. https://doi.org/10.1016/j.jmatprotec.2007.10.024

Chiang, K. T., & Chang, F. P. (2006). Optimization of the WEDM process of particle-reinforced material with multiple performance characteristics using grey relational analysis. *Journal of Materials Processing Technology, 180*(1–3), 96–101. https://doi.org/10.1016/j.jmatprotec.2006.05.008

Cho, S., Asfour, S., Onar, A., & Kaundinya, N. (2005). Tool breakage detection using support vector machine learning in a milling process. *International Journal of Machine Tools and Manufacture, 45*(3), 241–249. https://doi.org/10.1016/j.ijmachtools.2004.08.016

Ćojbašić, Ž., Petković, D., Shamshirband, S., Tong, C. W., Ch, S., Janković, P., Dučić, N., & Baralić, J. (2016). Surface roughness prediction by extreme learning machine constructed with abrasive water jet. *Precision Engineering, 43*, 86–92. https://doi.org/10.1016/j.precisioneng.2015.06.013

D'Addona, D. M., Ullah, A. M. M. S., & Matarazzo, D. (2017). Tool-wear prediction and pattern-recognition using artificial neural network and DNA-based computing. *Journal of Intelligent Manufacturing, 28*(6), 1285–1301. https://doi.org/10.1007/s10845-015-1155-0

Dalenogare, L. S., Benitez, G. B., Ayala, N. F., & Frank, A. G. (2018). The expected contribution of Industry 4.0 technologies for industrial performance. *International Journal of Production Economics, 204*(August), 383–394. https://doi.org/10.1016/j.ijpe.2018.08.019

Do, T. N., Lenca, P., Lallich, S., Pham, N. K. (2010). Classifying very-high-dimensional data with random forests of oblique decision trees. In: Guillet, F., Ritschard, G., Zighed, D. A., Briand, H. (Eds.), *Advances in Knowledge Discovery and Management. Studies in Computational Intelligence*, Vol 292. Springer. https://doi.org/10.1007/978-3-642-00580-0_3

Doltsinis, S., Ferreira, P., & Lohse, N. (2018). A symbiotic human-machine learning approach for production ramp-up. *IEEE Transactions on Human-Machine Systems, 48*(3), 229–240. https://doi.org/10.1109/THMS.2017.2717885

Elangovan, M., Sakthivel, N. R., Saravanamurugan, S., Nair, B. B., & Sugumaran, V. (2015). Machine learning approach to the prediction of surface roughness using statistical features of vibration signal acquired in turning. *Procedia Computer Science, 50*, 282–288. https://doi.org/10.1016/j.procs.2015.04.047

Filipič, B., & Junkar, M. (2000). Using inductive machine learning to support decision making in machining processes. *Computers in Industry*, 43(1), 31–41. https://doi.org/10.1016/S0166-3615(00)00056-7

Gibson, I., Rosen, D., & Stucker, B. (2015). 3D printing, rapid prototyping, and direct digital manufacturing. In *Additive Manufacturing Technologies*, Second edition. http://link.springer.com/10.1007/978-1-4939-2113-3

Guo, X., Sun, L., Li, G., & Wang, S. (2008). A hybrid wavelet analysis and support vector machines in forecasting development of manufacturing. *Expert Systems with Applications*, 35(1–2), 415–422. https://doi.org/10.1016/j.eswa.2007.07.052

Harding, J. A., Shahbaz, M., Srinivas, & Kusiak, A. (2006). Data mining in manufacturing: A review. *Journal of Manufacturing Science and Engineering*, 128(4), 969–976. https://doi.org/10.1115/1.2194554

Hofmann, E., & Rüsch, M. (2017). Industry 4.0 and the current status as well as future prospects on logistics. *Computers in Industry*, 89, 23–34. https://doi.org/10.1016/j.compind.2017.04.002

Jeschke, S., Brecher, C., Meisen, T., Özdemir, D., & Eschert, T. (2017, October, 3–19). Cyber-physical systems engineering for manufacturing in industrial internet of things. *Industrial Internet of Things. Springer Series in Wireless Technology*. Springer, Cham. https://doi.org/10.1007/978-3-319-42559-7

Jurkovic, Z., Cukor, G., Brezocnik, M., & Brajkovic, T. (2018). A comparison of machine learning methods for cutting parameters prediction in high speed turning process. *Journal of Intelligent Manufacturing*, 29(8), 1683–1693. https://doi.org/10.1007/s10845-016-1206-1

Rao, R. V., & Kalyankar, V. D. (2011, October). Parameters optimization of advanced machining processes using TLBO algorithm. *EPPM, Singapore*, 20, 21–32. https://doi.org/10.32738/ceppm.201109.0003

Karam, S., Centobelli, P., D'Addona, D. M., & Teti, R. (2016). Online prediction of cutting tool life in turning via cognitive decision making. *Procedia CIRP*, 41, 927–932. https://doi.org/10.1016/j.procir.2016.01.002

Khan, A. I., & Al-Badi, A. (2020). Open Source machine learning frameworks for industrial internet of things. *Procedia Computer Science*, 170, 571–577. https://doi.org/10.1016/j.procs.2020.03.127

Kim, D., Kang, P., Cho, S., Lee, H. J., & Doh, S. (2012). Machine learning-based novelty detection for faulty wafer detection in semiconductor manufacturing. *Expert Systems with Applications*, 39(4), 4075–4083. https://doi.org/10.1016/j.eswa.2011.09.088

Kim, D. H., Kim, T. J. Y., Wang, X., Kim, M., Quan, Y. J., Oh, J. W., Min, S. H., Kim, H., Bhandari, B., Yang, I., & Ahn, S. H. (2018). Smart machining process using machine learning: A review and perspective on machining industry. *International Journal of Precision Engineering and Manufacturing - Green Technology*, 5(4), 555–568. https://doi.org/10.1007/s40684-018-0057-y

Köksal, G., Batmaz, I., & Testik, M. C. (2011). A review of data mining applications for quality improvement in manufacturing industry. *Expert Systems with Applications*, 38(10), 13448–13467. https://doi.org/10.1016/j.eswa.2011.04.063

Konieczny, R., & Idczak, R. (2016). Mössbauer study of Fe-Re alloys prepared by mechanical alloying. *Hyperfine Interactions*, 237(1), 1–8. https://doi.org/10.1007/s10751-016-1232-6

Krishnakumar, P., Rameshkumar, K., & Ramachandran, K. I. (2015). Tool wear condition prediction using vibration signals in high speed machining (HSM) of Titanium (Ti-6Al-4V) alloy. *Procedia Computer Science, 50,* 270–275. https://doi.org/10.1016/j.procs.2015.04.049

Kubat, M. (2017). An introduction to machine learning. In *An Introduction to Machine Learning,* Second edition, Springer text book, https://doi.org/10.1007/978-3-319-63913-0

Kwak, D. S., & Kim, K. J. (2012). A data mining approach considering missing values for the optimization of semiconductor-manufacturing processes. *Expert Systems with Applications, 39*(3), 2590–2596. https://doi.org/10.1016/j.eswa.2011.08.114

Laha, D., Ren, Y., & Suganthan, P. N. (2015). Modeling of steelmaking process with effective machine learning techniques. *Expert Systems with Applications, 42*(10), 4687–4696. https://doi.org/10.1016/j.eswa.2015.01.030

Learning, M. (2010). Encyclopedia of machine learning. In *Encyclopedia of Machine Learning,* Springer, Springer, referance work. https://doi.org/10.1007/978-0-387-30164-8

Lee, J. H., & Ha, S. H. (2009). Recognizing yield patterns through hybrid applications of machine learning techniques. *Information Sciences, 179*(6), 844–850. https://doi.org/10.1016/j.ins.2008.11.008

Liao, Y., Deschamps, F., Loures, E. de F. R., & Ramos, L. F. P. (2017). Past, present and future of Industry 4.0 - a systematic literature review and research agenda proposal. *International Journal of Production Research, 55*(12), 3609–3629. https://doi.org/10.1080/00207543.2017.1308576

Lin, W., Yu, D. Y., Wang, S., Zhang, C., Zhang, S., Tian, H., Luo, M., & Liu, S. (2015). Multi-objective teaching-learning-based optimization algorithm for reducing carbon emissions and operation time in turning operations. *Engineering Optimization, 47*(7), 994–1007. https://doi.org/10.1080/0305215X.2014.928818

Lu, S. C. Y. (1990). Machine learning approaches to knowledge synthesis and integration tasks for advanced engineering automation. *Computers in Industry, 15*(1–2), 105–120. https://doi.org/10.1016/0166-3615(90)90088-7

Lu, X., Hu, X., Wang, H., Si, L., Liu, Y., & Gao, L. (2016). Research on the prediction model of micro-milling surface roughness of Inconel718 based on SVM. *Industrial Lubrication and Tribology, 68*(2), 206–211. https://doi.org/10.1108/ILT-06-2015-0079

Majumder, A. (2015). Comparative study of three evolutionary algorithms coupled with neural network model for optimization of electric discharge machining process parameters. *Proceedings of the Institution of Mechanical Engineers, Part B: Journal of Engineering Manufacture, 229*(9), 1504–1516. https://doi.org/10.1177/0954405414538960

Mat Deris, A., Mohd Zain, A., & Sallehuddin, R. (2013). Hybrid GR-SVM for prediction of surface roughness in abrasive water jet machining. *Meccanica, 48*(8), 1937–1945. https://doi.org/10.1007/s11012-013-9710-2

Mckinsey. (2018). Crossing the frontier: How to apply AI for impact. *McKinsey Analytics,* June, 111. Mckinsey and company. https://www.mckinsey.com/~/media/McKinsey/BusinessFunctions/McKinseyAnalytics/OurInsights/CrossingthefrontierHowtoapplyAIforimpact/Crossing-the-frontier-collection.ashx

Mekid, S., Pruschek, P., & Hernandez, J. (2009). Beyond intelligent manufacturing: A new generation of flexible intelligent NC machines. *Mechanism and Machine Theory*, 44(2), 466–476. https://doi.org/10.1016/j.mechmachtheory.2008.03.006

Monostori. (1993). A step towards intelligent manufacturing Modelling and monitoring of manufacturing processes through artificial neural networks. *CIRP Annals*, 42(1), 485–488. https://doi.org/10.1016/S0007-8506(07)62491-3

Nilsson, N. J. (2005). Introduction to machine learning an early draft of a proposed textbook, department of computer science. *Machine Learning*, 56(2), 387–399. http://www.ncbi.nlm.nih.gov/pubmed/21172442

Nti, I. K., Adekoya, A. F., Weyori, B. A., & Nyarko-Boateng, O. (2022). Applications of artificial intelligence in engineering and manufacturing: a systematic review. *Journal of Intelligent Manufacturing*, 33(6), 1581–1601. https://doi.org/10.1007/s10845-021-01771-6

Painuli, S., Elangovan, M., & Sugumaran, V. (2014). Tool condition monitoring using K-star algorithm. *Expert Systems with Applications*, 41(6), 2638–2643. https://doi.org/10.1016/j.eswa.2013.11.005

Park, J., Law, K. H., Bhinge, R., Biswas, N., Srinivasan, A., Dornfeld, D. A., Helu, M., & Rachuri, S. (2015). A generalized data-driven energy prediction model with uncertainty for a milling machine tool using Gaussian Process. *ASME 2015 International Manufacturing Science and Engineering Conference, MSEC 2015, 2*, 1–10. https://doi.org/10.1115/MSEC20159354

Pham, D. T., & Afify, A. A. (2005). Machine-learning techniques and their applications in manufacturing. *Proceedings of the Institution of Mechanical Engineers, Part B: Journal of Engineering Manufacture*, 219(5), 395–412. https://doi.org/10.1243/095440505X32274

Posada, J., Toro, C., Barandiaran, I., Oyarzun, D., Stricker, D., De Amicis, R., Pinto, E. B., Eisert, P., Döllner, J., & Vallarino, I. (2015). Visual computing as a key enabling technology for Industrie 4.0 and industrial internet. *IEEE Computer Graphics and Applications*, 35(2), 26–40. https://doi.org/10.1109/MCG.2015.45

Priore, P., De La Fuente, D., Gomez, A., & Puente, J. (2001). A review of machine learning in dynamic scheduling of flexible manufacturing systems. In *Artificial Intelligence for Engineering Design, Analysis and Manufacturing: AIEDAM* (Vol. 15, Issue 3). https://doi.org/10.1017/S0890060401153059

Rai, R., Tiwari, M. K., Ivanov, D., & Dolgui, A. (2021). Machine learning in manufacturing and industry 4.0 applications. *International Journal of Production Research*, 59(16), 4773–4778. https://doi.org/10.1080/00207543.2021.1956675

Saravanamurugan, S., Thiyagu, S., Sakthivel, N. R., & Nair, B. B. (2017). Chatter prediction in boring process using machine learning technique. *International Journal of Manufacturing Research*, 12(4), 405–422. https://doi.org/10.1504/IJMR.2017.088399

Schwab, K. (2018). The fourth industrial revolution (Industry 4.0) a social innovation perspective. *Tạp Chí Nghiên Cứu Dân Tộc*, 7(23), 12–21. https://doi.org/10.25073/0866-773x/97

Shaban, Y., Yacout, S., Balazinski, M., Meshreki, M., & Attia, H. (2015). Diagnosis of machining outcomes based on machine learning with Logical Analysis of Data. *IEOM 2015 - 5th International Conference on Industrial Engineering and Operations Management, Proceeding*, May. https://doi.org/10.1109/IEOM.2015.7093752

Sharma, R., Jabbour, C. J. C., & Lopes de Sousa Jabbour, A. B. (2020). Sustainable manufacturing and industry 4.0: What we know and what we don't. *Journal of Enterprise Information Management*, *34*(1), 230–266. https://doi.org/10.1108/JEIM-01-2020-0024

Somashekhar, K. P., Ramachandran, N., & Mathew, J. (2010). Optimization of material removal rate in micro-EDM using artificial neural network and genetic algorithms. *Materials and Manufacturing Processes*, *25*(6), 467–475. https://doi.org/10.1080/10426910903365760

Susto, G. A., Schirru, A., Pampuri, S., McLoone, S., & Beghi, A. (2015). Machine learning for predictive maintenance: A multiple classifier approach. *IEEE Transactions on Industrial Informatics*, *11*(3), 812–820. https://doi.org/10.1109/TII.2014.2349359

Teixidor, D., Grzenda, M., Bustillo, A., & Ciurana, J. (2015). Modeling pulsed laser micromachining of micro geometries using machine-learning techniques. *Journal of Intelligent Manufacturing*, *26*(4), 801–814. https://doi.org/10.1007/s10845-013-0835-x

Tosun, N., & Özler, L. (2002). A study of tool life in hot machining using artificial neural networks and regression analysis method. *Journal of Materials Processing Technology*, *124*(1–2), 99–104. https://doi.org/10.1016/S0924-0136(02)00086-9

Wan, J. (2015). *Industrie 4.0: Enabling Technologies*. 135–140, 2015 *International Conference on Intelligent Computing and Internet of Things (IC1T)*. https://doi.org/10.1109/ICAIOT.2015.7111555

Wang, L. (2019). From Intelligence Science to Intelligent Manufacturing. *Engineering*, *5*(4), 615–618. https://doi.org/10.1016/j.eng.2019.04.011

Wang, Y., Ma, H. S., Yang, J. H., & Wang, K. S. (2017). Industry 4.0: a way from mass customization to mass personalization production. *Advances in Manufacturing*, *5*(4), 311–320. https://doi.org/10.1007/s40436-017-0204-7

Widodo, A., & Yang, B. S. (2007). Support vector machine in machine condition monitoring and fault diagnosis. *Mechanical Systems and Signal Processing*, *21*(6), 2560–2574. https://doi.org/10.1016/j.ymssp.2006.12.007

Wu, D., Jennings, C., Terpenny, J., Gao, R. X., & Kumara, S. (2017). A comparative study on machine learning algorithms for smart manufacturing: Tool wear prediction using random forests. *Journal of Manufacturing Science and Engineering, Transactions of the ASME*, *139*(7), 1–9. https://doi.org/10.1115/1.4036350

Wuest, T., Weimer, D., Irgens, C., & Thoben, K. D. (2016). Machine learning in manufacturing: Advantages, challenges, and applications. *Production and Manufacturing Research*, *4*(1), 23–45. https://doi.org/10.1080/21693277.2016.1192517

Yu, L., & Liu, H. (2003, August 21–24). Feature selection for high-dimensional data: A fast correlation-based filter solution. *Conference: Machine Learning, Proceedings of the Twentieth International Conference (ICML 2003)*, Washington, DC, USA.

Yuan, J., Wang, K., Yu, T., & Fang, M. (2008). Reliable multi-objective optimization of high-speed WEDM process based on Gaussian process regression. *International Journal of Machine Tools and Manufacture*, *48*(1), 47–60. https://doi.org/10.1016/j.ijmachtools.2007.07.011

Zhang, D., Bi, G., Sun, Z., & Guo, Y. (2015). Online monitoring of precision optics grinding using acoustic emission based on support vector machine. *International Journal of Advanced Manufacturing Technology*, *80*(5–8), 761–774. https://doi.org/10.1007/s00170-015-7029-y

Index

Pages in *italics* refer to figures and pages in **bold** refer to tables.

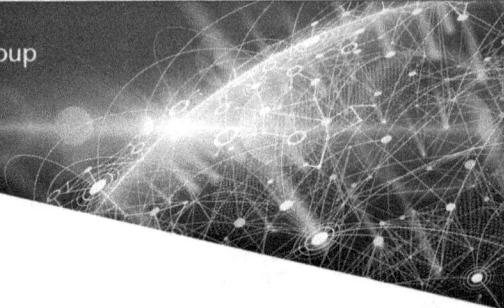

For Product Safety Concerns and Information please contact our EU
representative GPSR@taylorandfrancis.com
Taylor & Francis Verlag GmbH, Kaufingerstraße 24, 80331 München, Germany

www.ingramcontent.com/pod-product-compliance
Lightning Source LLC
Chambersburg PA
CBHW060359220326
41598CB00023B/2970